Probability and Mathemati

SEBER • Linear Reg
SEBER • Multivariat
SEN • Sequential No nd Statistical Inference
SERFLING • Approximation Theorems of Mathematical Statistics
TJUR • Probability Based on Radon Measures
WILLIAMS • Diffusions, Markov Processes, and Martingales, Volume I: Foundations
ZACKS • Theory of Statistical Inference

Applied Probability and Statistics

ABRAHAM and LEDOLTER • Statistical Methods for Forecasting
AGRESTI • Analysis of Ordinal Categorical Data
AICKIN • Linear Statistical Analysis of Discrete Data
ANDERSON, AUQUIER, HAUCK, OAKES, VANDAELE, and WEISBERG • Statistical Methods for Comparative Studies
ARTHANARI and DODGE • Mathematical Programming in Statistics
BAILEY • The Elements of Stochastic Processes with Applications to the Natural Sciences
BAILEY • Mathematics, Statistics and Systems for Health
BARNETT • Interpreting Multivariate Data
BARNETT and LEWIS • Outliers in Statistical Data
BARTHOLOMEW • Stochastic Models for Social Processes, *Third Edition*
BARTHOLOMEW and FORBES • Statistical Techniques for Manpower Planning
BECK and ARNOLD • Parameter Estimation in Engineering and Science
BELSLEY, KUH, and WELSCH • Regression Diagnostics: Identifying Influential Data and Sources of Collinearity
BHAT • Elements of Applied Stochastic Processes
BLOOMFIELD • Fourier Analysis of Time Series: An Introduction
BOX • R. A. Fisher, The Life of a Scientist
BOX and DRAPER • Evolutionary Operation: A Statistical Method for Process Improvement
BOX, HUNTER, and HUNTER • Statistics for Experimenters: An Introduction to Design, Data Analysis, and Model Building
BROWN and HOLLANDER • Statistics: A Biomedical Introduction
BROWNLEE • Statistical Theory and Methodology in Science and Engineering, *Second Edition*
CHAMBERS • Computational Methods for Data Analysis
CHATTERJEE and PRICE • Regression Analysis by Example
CHOW • Analysis and Control of Dynamic Economic Systems
CHOW • Econometric Analysis by Control Methods
COCHRAN • Sampling Techniques, *Third Edition*
COCHRAN and COX • Experimental Designs, *Second Edition*
CONOVER • Practical Nonparametric Statistics, *Second Edition*
CONOVER and IMAN • Introduction to Modern Business Statistics
CORNELL • Experiments with Mixtures: Designs, Models and The Analysis of Mixture Data
COX • Planning of Experiments
DANIEL • Biostatistics: A Foundation for Analysis in the Health Sciences, *Third Edition*
DANIEL • Applications of Statistics to Industrial Experimentation
DANIEL and WOOD • Fitting Equations to Data: Computer Analysis of Multifactor Data, *Second Edition*
DAVID • Order Statistics, *Second Edition*
DAVISON • Multidimensional Scaling
DEMING • Sample Design in Business Research
DILLON and GOLDSTEIN • Multivariate Analysis: Methods and Applications
DODGE and ROMIG • Sampling Inspection Tables, *Second Edition*
DOWDY and WEARDEN • Statistics for Research

continued on back

ANALYSIS OF ORDINAL CATEGORICAL DATA

ANALYSIS OF ORDINAL CATEGORICAL DATA

ALAN AGRESTI
University of Florida

JOHN WILEY & SONS

New York • Chichester • Brisbane • Toronto • Singapore

Library of Congress Cataloging in Publication Data:

Agresti, Alan.
Analysis of ordinal categorical data.

(Wiley series in probability and mathematical statistics. Applied probability and statistics)
Bibliography: p.
Includes index.
1. Multivariate analysis. I. Title. II. Series.

QA278.A35 1984 519.5′35 83-23535
ISBN 0-471-89055-3

Printed in the United States of America

10 9 8 7 6 5 4 3 2

Preface

In the past decade methods for analyzing cross-classification tables have received considerable attention. There has been a tremendous increase in the publication of articles in the statistics literature on this topic. Several books have introduced the methods to audiences of nonstatisticians as well as statisticians, and the methods are now frequently applied by researchers in areas as diverse as sociology, public health, ecology, and pharmacy.

What distinguishes this book from others recently published is its emphasis on methods for cross-classification tables having ordered categories for at least one of the classifications. Specialized models and descriptive measures are discussed that, unlike standard methods for "nominal" categorical data, efficiently use the information on the ordering. The "ordinal" methods make possible simpler description of the data and also permit more powerful inferences about population characteristics. In particular, this book presents detailed descriptions of loglinear and logit ordinal models that have been proposed within the past few years.

For completeness, I have included an introduction to basic descriptive and inferential methods for categorical data (ordinal *or* nominal) in Chapters 2 through 4. Readers who are familiar with chi-squared tests and standard loglinear models for multidimensional tables may wish to skip to Chapter 5, where the presentation of methods pertaining solely to ordinal data begins.

The overwhelming majority of readers will probably use these ordinal methods by applying existing computer software, rather than by writing their own programs for model fitting and parameter estimation. Thus I have placed more emphasis on interpretation and application of the methods than on technical details regarding estimation techniques. For interested readers these technical details are outlined in Appendixes A through C. Appendix D gives brief descriptions of several computer packages and specialized programs for applying the methods discussed in this book. Other supplementary technical comments and embellishments are given in the sections headed Notes and Complements (exercises) at the ends of the chapters.

The analyses in the text of several data sets illustrate to the reader the power, yet simplicity, of ordinal methods. A few of these data sets are analyzed several ways throughout the text so that the reader can compare and contrast alternative strategies.

I intend this book to be accessible to a broad audience—particularly professional statisticians and researchers in areas such as public health, the pharmaceutical industry, the social and behavioral sciences, and business and government. The reader should possess a basic background in statistical methods, especially including regression and analysis of variance models. Because of the inclusion of the introductory chapters, previous exposure to categorical data methodology is unnecessary. Readers who want further details on standard methods for categorical data are referred to the compendium by Bishop, Fienberg, and Holland (1975).

This book can also be used as a text in a special topics course or as a supplementary text in courses on categorical data analysis. The notes that gave birth to this book have been used in this way at the University of Florida and at Birkbeck College, University of London.

I would like to thank those individuals who commented on parts of the manuscript. These include Prof. D. R. Cox, Dr. B. Jørgensen, and three anonymous reviewers. Also thanks to Dr. C. C. Clogg and to Dr. P. McCullagh for providing me with personal computer programs. Most of the work on this book was completed while I was visiting the Mathematics Department at Imperial College, London, on sabbatical leave. I thank the staff there for their kind hospitality. Finally, I would like to dedicate this book to my English and Italian relatives and to the many wonderful friends I had the chance to make during my sabbatical year.

ALAN AGRESTI

Gainesville, Florida
February 1984

Contents

ANALYSIS OF ORDINAL CATEGORICAL DATA

CHAPTER 1

Introduction

1.1. ORDINAL CATEGORICAL SCALES

In the past decade methods for the analysis of ordinal categorical data have become increasingly well developed. This is reflected by the large number of books published on the topic during that period, for example, those by Bishop, Fienberg, and Holland (1975), Cox (1970), Everitt (1977), Fienberg (1980), Fleiss (1981), Forthofer and Lehnen (1981), Gilbert (1981), Gokhale and Kullback (1978), Goodman (1978), Haberman (1974a, 1978, 1979), Knoke and Burke (1980), Plackett (1981), Reynolds (1977), and Upton (1978).

What distinguishes this book from these others is its emphasis on methods that can (and should) be used when one or more of the variables is measured on an *ordinal* scale. These scales consist of a collection of naturally ordered categories. As an illustration of ordinal variables and the levels of their corresponding scales, political philosophy may be classified as "liberal," "moderate," or "conservative"; social class may be measured as "upper," "middle," or "lower"; opinion on abortion might have responses such as "should be available on demand," "should only be allowed in particular circumstances," or "should never be allowed."

As can be seen from these examples, ordinal scales are pervasive in the social sciences, in particular for measuring attitudes and opinions on various issues and status of various types. Ordinal scales also commonly occur in such diverse fields as marketing (e.g., ordinal preference scales or resource scales) and medical and public health disciplines (e.g., for variables describing severity of an injury, degree of recovery from an illness, stages of a disease, and amount of exposure to a potentially harmful substance). In all fields ordinal scales often result when discrete measurement is used with inherently continuous variables such as age, education, and degree of prejudice.

A categorical variable is referred to as "ordinal" rather than "interval" when there is a clear ordering of the categories but the absolute distances among them are unknown. For example, the variable "education" is ordinal when measured with categories grammar school, high school, college, post-graduate, but it is interval when measured with the integer values 0, 1, 2, ... representing number of years of education. Political philosophy is ordinal, because a person categorized as moderate is *more* liberal than a person categorized as conservative, but there is no obvious way to quantify numerically *how much more* liberal that person is.

Two points should be made at this stage. First, inherently continuous interval variables such as number of years of education are often measured in an ordinal manner by categorizing into groupings such as 0–8, 9–12, 13–16, 17 or more. Hence the methods discussed in this book may be more appropriate for them than the parametric and nonparametric techniques that assume continuous distributions. Conversely, for many ordinal categorical variables it is sensible to imagine the existence of an underlying continuous variable. To approximate the underlying scale, it is often useful to assign a "reasonable" set of scores to the categories. Some of the methods discussed in this text require this to be done. Unfortunately, strict adherence to operations that utilize only the ordering in ordinal scales limits too severely the scope of useful methodology. This is particularly true for model building and for multivariate analyses. We therefore do not take a rigid view in this text regarding permissible methodology for ordinal variables, and we do consider some score-based methods. At worst, the robustness of substantive conclusions to the choice of scores can be checked. One can, for instance, assign scores in a variety of ways for those variables (e.g., opinion about abortion) where no obvious choice exists.

An ordinal (or interval) variable is *quantitative* in the sense that each level on its scale can be compared in terms of whether it corresponds to a *greater* or *smaller* magnitude of a certain characteristic than another level. Such variables are of a very different nature than *qualitative* variables, which are measured on a *nominal* scale. Examples of nominal variables are race, county of residence, religious affiliation, and marital status. Distinct levels of such variables differ in quality, not in quantity. Therefore, the order of listing of the categories of a nominal variable is unimportant.

1.2. ADVANTAGES OF USING ORDINAL METHODS

Most of the well-known statistical methods for analyzing categorical data treat all variables as nominal. That is, the results are invariant to permutations of the categories of any of the variables. Examples of these methods are the Pearson chi-squared test of independence (Section 2.2) and the traditional loglinear models (Chapters 3–4) that form the basis of books such as the one by Bishop et al. (1975). In much of the research conducted in various disciplines, these methods are routinely applied to nominal and ordinal data alike because they are both "categorical."

A recognition of the categorical nature of data *is* useful for formulating sampling models (e.g., multinomial instead of normal). However, the distinction regarding whether data are continuous or discrete is often less important than whether the data are qualitative (nominal) or quantitative (ordinal or interval). Since ordinal variables are inherently quantitative, their descriptive measures should be more like those for interval variables than those for nominal variables. For example, the models and measures of association for ordinal data presented in Chapters 5 through 12 of this book bear many resemblances to those for continuous variables.

Hence the major theme of this book is the importance of utilizing the quantitative nature of ordinal variables in the analysis of categorical data. In several examples we shall see that the *type* of ordinal method used is not that crucial, in the sense that similar substantive results are usually obtained whether we use loglinear models, logit models, models with other types of response functions, or measures of association and nonparametric procedures. These results may be quite different, however, from those obtained using standard categorical data techniques that treat all the variables as nominal.

There are many advantages to be gained from using ordinal methods instead of (or in addition to) the standard nominal procedures. These advantages will be discussed in detail in various parts of this book. In summary, they include the following:

1. Ordinal methods have greater power for detecting important alternatives to null hypotheses such as the one of independence.
2. Ordinal data description is based on measures that are similar to those (e.g., correlations, slopes, means) used in ordinary regression and analysis of variance for continuous variables.
3. Ordinal analyses can use a greater variety of models, most of which are more parsimonious and have simpler interpretations than the standard models for nominal variables.
4. Interesting ordinal models can be applied in settings where the standard nominal models are trivial or else have too many parameters to be tested for goodness of fit.

1.3. ORGANIZATION OF THIS BOOK

The primary methodological emphasis in this book is on constructing models that can describe associations and provide a framework for making inferences. Two types of models are presented in some detail. Loglinear models describe association patterns among all variables. They make no distinction between response (or "dependent") variables and explanatory (or "independent") variables. Logit models, on the other hand, describe effects of explanatory variables on a response variable. They closely resemble regression and analysis of variance models for continuous response variables. The loglinear and logit approaches form the core of the book and are discussed in Chapters 3 through 7 and 12.

Chapters 2 through 4 introduce basic results and concepts that apply to categorical data of nominal *or* ordinal nature. The specialized methods for ordinal variables then build on these general results. Hence it is not necessary for the reader to have prior familiarity with any form of categorical data analysis to be able to read this book.

Chapter 2 introduces much of the terminology used in categorical data analysis and provides a brief introduction to chi-squared tests and measures of association. Chapter 3 gives an introduction to the types of association patterns that can exist in a multivariate setting. Chapter 4 describes the fitting of loglinear models that represent these patterns, and Chapter 5 introduces loglinear models for cases where at least one variable is ordinal. Chapter 6 introduces logit models, and their use with ordinal response variables is described in Chapter 7. Other types of models that can be applied to ordinal data are summarized in Chapter 8. Chapters 9 and 10 have a somewhat different, less parametric, emphasis than the others. There, descriptive and inferential procedures are based on summary measures of association that need not be connected with structural models. Specialized methods for square tables having ordered categories are presented in Chapter 11. Chapter 12 summarizes and compares the various strategies for analyzing ordinal categorical data. Finally in the appendixes further details are given on maximum likelihood and weighted least squares estimation for ordinal models, asymptotic, distribution theory, and computer programs for fitting ordinal models.

CHAPTER 2

Basic Results for Cross-Classification Tables

This chapter introduces some basic terminology and some basic association statistics for categorical data. The methods discussed are designed for bivariate analyses. They are appropriate, for instance, in assessing the association between race and employment status, or between marital status and whether an alcoholic. Chapters 3 and 4 introduce basic multivariate analyses for categorical data. Most of the methodology in Chapters 2 through 4 can be applied to nominal or ordinal variables.

This chapter is intended as a brief summary for readers who may not have any prior familiarity with methodology for categorical data. Readers who already have this background may wish to study the material on odds ratios for ordinal data in Section 2.4 and then proceed to Chapters 5 through 12 for specialized methods for ordinal variables.

2.1. CROSS-CLASSIFICATION TABLES

This book contains analyses that are appropriate when each member of a sample is classified simultaneously on two or more categorical variables. The analyses are applied to a table that displays the frequencies of observations occurring at the various combinations of levels of the variables. Each cell in the table gives a count of the number of observations that have a certain combination of characteristics—such as unmarried alcoholics. The table is called a *cross-classification table*, and is also referred to as a "contingency table" or "cross-tab." Tables for two variables have two dimensions, r rows that represent the categories of one variable and c columns that represent the categories of a second variable. The rc cells of the table contain frequencies of occurrence of the rc combinations of categories of the two variables.

Table 2.1. Death Penalty Verdict by Defendant's Race

	Death Penalty	
Defendant's Race	Yes	No
White	19	141
Black	17	149

Source: Radelet (1981).

An important special case of the $r \times c$ cross-classification table occurs for $r = c = 2$. The 2×2 table commonly results from comparisons of two groups on a dichotomous (two category, or "binary") response; for example, a comparison of men and women with respect to the numbers who favor and who oppose increased defense spending, or a comparison of two drugs on their frequencies of positive and negative effects.

Death Penalty Example

Table 2.1, based on data presented by Radelet (1981), is an example of a 2×2 cross-classification table. Radelet's article concerns the effects of racial characteristics on the decision regarding whether to impose the death penalty after an individual is convicted for a homicide. The variables considered in Table 2.1 are "death penalty verdict," having categories (yes, no), and "race of defendant," having categories (white, black). The 326 subjects cross-classified according to these variables were defendants in homicide indictments in 20 Florida counties during 1976–1977. Table 2.1 refers only to indictments for homicides in which the defendant and the victim were strangers, since death sentences are very rarely imposed when the defendant and the victim had a prior friendship or family relationship.

Notation and Definitions

Let n_{ij} denote the number of observations cross-classified in the cell of the table that is in row i and column j, and let p_{ij} denote the proportion of the total sample falling in that cell. That is $p_{ij} = n_{ij}/n$, where $n = \sum_i \sum_j n_{ij}$ is the total sample size, so that $\sum_i \sum_j p_{ij} = 1.0$. The set $\{p_{ij}\}$ is called the sample *joint distribution*. The sample *marginal distributions* are the row totals and column totals obtained by summing the joint proportions. These will be denoted by $\{p_{i+}\}$ for the row variable, where $p_{i+} = \sum_j p_{ij}$, and by $\{p_{+j}\}$ for the column variable, where $p_{+j} = \sum_i p_{ij}$. Note that $p_{i+} = n_{i+}/n$ and $p_{+j} = n_{+j}/n$, and that $\sum_i p_{i+} = \sum_j p_{+j} = 1.0$. For Table 2.1, for instance, 36 defendants received the death penalty, and 290 defendants did not receive it, so the marginal distribution for death penalty verdict is $p_{+1} = 36/326 = 0.110$ and $p_{+2} = 290/326 = 0.890$.

In most applications it is natural to treat one of the variables in a cross-classification table as a response variable. The table then provides information regarding how the response on this variable depends on an individual's classification on the explanatory variable. Different levels of the explanatory variable can be compared with respect to the proportions of observations in the various categories of the response variable. The collection of response proportions at a certain level of the explanatory variable is called a sample *conditional distribution.*

For the data in Table 2.1, death penalty verdict is the response variable. The sample conditional distribution on the death penalty verdict for a given race of defendant can be obtained by dividing the cell frequencies by the sample size for that race. For the condition "defendant's race = white," the sample proportions are $19/160 = 0.119$ and $141/160 = 0.881$. For the condition "defendant's race = black," the sample proportions are 0.102 and 0.898. Hence 11.9 percent of the white defendants received the death penalty, and 10.2 percent of the black defendants received the death penalty.

This book follows the convention of forming cross-classification tables so that the levels of the response variable are the columns. For the sample conditional distribution in row i, we denote the proportion of the observations whose response is the jth category of the column variable by $p_{j(i)}$. Hence $p_{j(i)} = n_{ij}/n_{i+}$, where $n_{i+} = \sum_j n_{ij}$ is the ith row total and $\sum_j p_{j(i)} = 1.0$. For Table 2.1, for instance, $p_{1(1)} = 0.119$ and $p_{2(1)} = 0.881$, whereas $p_{1(2)} = 0.102$ and $p_{2(2)} = 0.898$.

Notation introduced so far refers to data in a sample. Similar notation will be used for population proportions, with the Greek letter π in place of p. For instance, population analogs will be denoted by $\{\pi_{ij}\}$ for the joint probabilities and $\{\pi_{j(i)}\}$ for the conditional probabilities. Table 2.2 illustrates the notation for the 2×2 case. The population conditional, joint, and marginal probabilities are related by $\pi_{j(i)} = \pi_{ij}/\pi_{i+}$, and they satisfy $\sum_i \sum_j \pi_{ij} = 1.0$ and $\sum_j \pi_{j(i)} = 1.0$ for $i = 1, \ldots, r$.

Descriptive measures for cross-classifications of ordinal variables are sometimes expressed in terms of distribution functions. Let $F_{ij} = \sum_{a \leq i} \sum_{b \leq j} \pi_{ab}$, $i = 1, \ldots, r$ and $j = 1, \ldots, c$, denote the joint distribution function in the population. That is, F_{ij} is the probability of classification in the first i rows and the first j columns. Let X denote the row variable, and let Y denote the column variable. The marginal distribution functions are denoted by $F_i^X = \sum_{a \leq i} \pi_{a+} = F_{ic}$ and $F_j^Y = \sum_{b \leq j} \pi_{+b} = F_{rj}$. The conditional distribution function $F_{j(i)} = \sum_{b \leq j} \pi_{b(i)}$ gives the probability of classification in one of the first j columns, given classification in row i.

Table 2.2. Notation for Joint, Conditional, and Marginal Probabilities

Row	Column		Total
	1	2	
1	π_{11}	π_{12}	π_{1+}
	$(\pi_{1(1)})$	$(\pi_{2(1)})$	(1.0)
2	π_{21}	π_{22}	π_{2+}
	$(\pi_{1(2)})$	$(\pi_{2(2)})$	(1.0)
Total	π_{+1}	π_{+2}	1.0

Two variables are *independent* if all the joint probabilities equal the product of the corresponding marginal probabilities; that is, if

$$\pi_{ij} = \pi_{i+}\pi_{+j} \quad \text{for} \quad i = 1, \ldots, r \quad \text{and} \quad j = 1, \ldots, c \tag{2.1}$$

Independence of two variables implies that the conditional distributions within the r rows are identical, since $\pi_{j(i)} = \pi_{ij}/\pi_{i+} = (\pi_{i+}\pi_{+j})/\pi_{i+} = \pi_{+j}$ for $i = 1, \ldots, r$. Hence, if two variables are independent, the probability of making a particular response j on the column variable is the same in each row. In the next section we will discuss statistics that can be applied to sample data to test the null hypothesis of independence.

2.2. CHI-SQUARED STATISTICS

In analyzing cross-classification tables, we usually wish to use sample cell counts $\{n_{ij}\}$ not only for description but also to make inferences about the underlying structure of the table specified by unknown cell probabilities $\{\pi_{ij}\}$. In this section we introduce two statistics that are used for testing hypotheses about associations.

Multinomial Sampling

Suppose that a simple random sample of fixed size n is cross-classified according to the categorical variables. The a priori distribution of the cell counts is then the multinomial distribution specified by the sample size n and the rc population cell probabilities $\{\pi_{ij}\}$. The probability of a particular set of cell counts $\{n_{ij}\}$ that sum to n equals

$$\left(\frac{n!}{\prod_i \prod_j n_{ij}!}\right) \prod_i \prod_j \pi_{ij}^{n_{ij}}$$

This sampling scheme is called *full multinomial sampling.*

Suppose that within each category of the row variable, an independent random sample is classified according to the column variable. The cell counts within the ith row then have the multinomial distribution specified by the sample size n_{i+} and the response probabilities $\{\pi_{j(i)}, j = 1, \ldots, c\}$, and cell counts from different rows are independent. The cell counts in row i have the probability function

$$\left(\frac{n_{i+}!}{\prod_j n_{ij}!}\right) \prod_j \pi_{j(i)}^{n_{ij}}$$

and the product of these from the r rows gives the probability function for the entire table. This sampling scheme is called *independent* (or *product*) *multinomial sampling.*

Yet another sampling scheme assumes that the cell counts are independent Poisson random variables. If the cell count n_{ij} has expected value m_{ij}, then the probability function for n_{ij} has the form

$$\frac{e^{-m_{ij}} m_{ij}^{n_{ij}}}{n_{ij}!}$$

for all nonnegative integers. The product of these probabilities for the rc cells gives the probability function for the entire table. For Poisson sampling even the total sample size in the table is random. It can be shown that conditioning on the total sample size yields full multinomial sampling. Conditioning on the marginal totals in one margin yields independent multinomial sampling.

In the discussion here and in most of the text we assume the full multinomial scheme. For the statistics considered in this section, there is the nice occurrence of the same asymptotic distribution for all three sampling schemes. For full multinomial sampling, the maximum likelihood (ML) estimates of the marginal row proportions $\{\pi_{i+}\}$ and the marginal column proportions $\{\pi_{+j}\}$ are the corresponding sample proportions $\{p_{i+}\}$ and $\{p_{+j}\}$, respectively. Under the assumption of independence of the two variables, $\pi_{ij} = \pi_{i+}\pi_{+j}$ and the ML estimate of π_{ij} (denoted by $\hat{\pi}_{ij}$) is

$$\hat{\pi}_{ij} = p_{i+}p_{+j} = \frac{n_{i+}n_{+j}}{n^2} \tag{2.2}$$

For multinomial as well as Poisson sampling, we shall let m_{ij} represent the expected value of the cell count n_{ij}. If the $\{n_{ij}\}$ have a multinomial distribution, then each cell count has a binomial distribution. For full multinomial sampling, for instance, n_{ij} has the binomial distribution based on sample size n and "success probability" π_{ij}. Hence $m_{ij} = E(n_{ij}) = n\pi_{ij}$. The $\{m_{ij}\}$ are referred to as *expected frequencies.*

The estimate of m_{ij} is $\hat{m}_{ij} = n\hat{\pi}_{ij}$. Under the assumption of independence this equals

$$\hat{m}_{ij} = np_{i+}p_{+j} = \frac{n_{i+}n_{+j}}{n} \tag{2.3}$$

The $\{\hat{m}_{ij}\}$ are *estimated expected frequencies* for testing the null hypothesis of independence. The $\{\hat{m}_{ij}\}$ have the same marginal totals as do the observed data. For instance, $\hat{m}_{i+} = \sum_j \hat{m}_{ij} = (n_{i+}/n) \sum_j n_{+j} = n_{i+}$.

Pearson and Likelihood-Ratio Statistics

In 1900 Karl Pearson suggested the statistic

$$X^2 = \frac{\sum_i \sum_j (n_{ij} - \hat{m}_{ij})^2}{\hat{m}_{ij}} \tag{2.4}$$

for testing the null hypothesis H_0: independence.* The likelihood-ratio approach to testing hypotheses (see, e.g., Lindgren 1976, p. 437) leads to the test statistic

$$G^2 = 2 \sum_i \sum_j n_{ij} \log\left(\frac{n_{ij}}{\hat{m}_{ij}}\right) \tag{2.5}$$

When H_0: independence is true, both X^2 and G^2 have asymptotic (as $n \to \infty$) chi-squared distributions with degrees of freedom $\text{df} = (r - 1)(c - 1)$. In fact the two test statistics are asymptotically equivalent in that case, in the sense that the difference between the two is of smaller order than $1/n$ as $n \to \infty$. For either statistic larger values provide more evidence against the null hypothesis. Thus the attained significance level (P-value) is the right-hand tail probability of getting a statistic value larger than the observed one, assuming H_0 is true.

Various guidelines have been given for how large the sample size should be in order for the chi-squared distribution to give a good approximation for the exact sampling distributions of the X^2 and G^2 statistics. A commonly quoted guideline due to Cochran (1954) is that at least 80 percent of the cells should have $\hat{m}_{ij}$ exceeding 5.0, and $\hat{m}_{ij}$ should exceed 1.0 in all cells. Larntz (1978) and Koehler and Larntz (1980) showed that the chi-squared approximation can be very good for the X^2 statistic even for very small expected frequencies.

*We use the symbol X^2 to denote a statistic that has an approximate chi-squared distribution, and we use the symbol χ^2 to denote the distribution. For example, χ^2_6 denotes the chi-squared distribution with 6 degrees of freedom.

If many $\hat{m}_{ij}$ are very small, researchers commonly combine categories of variables to obtain a table having larger cell frequencies. Generally, one should not pool categories unless there is a natural way to combine them, and little information is lost in defining the variable more crudely. It is unnecessary to do so in any case, since there are alternatives to the chi-squared test when the cell counts are very small. An exact test can be conducted using a hypergeometric sampling distribution for cell frequencies. This test treats the marginal totals as given. For 2×2 tables the test is known as *Fisher's exact test* (see Note 2.4).

The $\{\hat{m}_{ij}\}$ in (2.3) depend on the row and column marginal totals, but not on the order in which the rows and columns are listed. Hence the X^2 and G^2 statistics for testing independence do not change under permutations of rows and permutations of columns. This implies that both classifications are treated as nominal scales in these tests. If we apply these statistics to testing independence between ordinal variables therefore, we are ignoring some of the available information. The consequences of this are considered in Sections 4.5 and 5.1.

Death Penalty Example

Table 2.3 contains the estimated expected frequencies for the death penalty data in parentheses below the corresponding observed counts. For example, the first value is $\hat{m}_{11} = n_{1+}n_{+1}/n = (160 \times 36)/326 = 17.67$. The chi-squared statistics equal $X^2 = 0.22$ and $G^2 = 0.22$, based on df = 1. The equality of values here is coincidental. Since the chi-squared distribution has mean equal to df, the observed values of X^2 and G^2 are not especially large. These data show no significant association between death penalty verdict and defendant's race.

Table 2.3. Observed and Estimated Expected Frequencies for Death Penalty Verdict and Defendant's Race

	Death Penalty		
Defendant's Race	Yes	No	Total
White	19	141	160
	(17.67)	(142.33)	
Black	17	149	166
	(18.33)	(147.67)	
Total	36	290	326

2.3. PARTITIONS OF CHI-SQUARED

If Z is a random variable having a standard normal distribution, then Z^2 has a chi-squared distribution with one degree of freedom. More generally, a chi-squared random variable with df $= d$ can be expressed as $\sum_{i=1}^{d} Z_i^2$, where $Z_1, \ldots, Z_d$ are independent standard normal random variables. From this construction follows one of the most useful properties of the chi-squared distribution, its "reproductive" nature. If X_1^2 and X_2^2 are independent random variables having chi-squared distributions with degrees of freedom df_1 and df_2, then $X^2 = X_1^2 + X_2^2$ has a chi-squared distribution with degrees of freedom $\text{df}_1 + \text{df}_2$. Conversely, a chi-squared statistic having df $= d > 1$ can be partitioned into independent chi-squared components, for example, into d components each having a single degree of freedom.

Chi-squared statistics can often be partitioned so that the components represent certain aspects of the overall association. Such a partitioning may show that the association in a table primarily reflects differences between certain rows or groupings of rows, or primarily reflects certain trends related to the ordering of the categories of an ordinal variable.

Partitioning Tables

We will illustrate one type of partitioning, due to Goodman (1968), using the data in Table 2.4. The data, originally presented by Grizzle, Starmer, and Koch (1969), refer to a comparison of four different operations for treating duodenal ulcer patients. The operations correspond to removal of various amounts of the stomach. Operation A is drainage and vagotomy, B is 25 percent resection and vagotomy, C is 50 percent resection and vagotomy, and D is 75 percent resection. The categories of operation have a

Table 2.4. Cross-Classification of Dumping Severity and Operation.

	Dumping Severity			
Operation	None	Slight	Moderate	Total
A	61	28	7	96
B	68	23	13	104
C	58	40	12	110
D	53	38	16	107
Total	240	129	48	417

Source: Grizzle, Starmer, and Koch (1969).

natural ordering, with A being the least severe operation and D corresponding to the greatest removal of stomach. The variable "dumping severity" describes the extent of a possible undesirable side effect of the operation. The categories of this variable are also ordered, with the response "none" representing the most desirable result.

The chi-squared statistics for testing independence for these data are $X^2 = 10.54$ and $G^2 = 10.88$, based on df = 6. From the table in Appendix E, these values correspond to attained significance levels of about $P = 0.10$. Hence there is some evidence of an association, but not enough to reject H_0 at common type I error rates such as $\alpha = 0.05$ or $\alpha = 0.01$.

Either of these statistics can be decomposed several ways into sums of chi-squared statistics. Here we form two chi-squared statistics, each having df = 3, as follows. One statistic compares the four operations for those individuals having some dumping. That is, a 4×2 table is formed from the last two columns of Table 2.4 in order to test whether the operations differ in relative numbers of slight and moderate dumping, for those outcomes where dumping occurred. For that table, $X^2 = 2.97$ and $G^2 = 2.96$ based on df = 3, indicating no evidence of a difference. The other statistic compares the four operations with respect to whether the amount of dumping is "none" or "some." In other words, a 4×2 table is formed by combining columns 2 and 3 (the ones previously compared) of Table 2.4. The test statistics $X^2 = 7.89$ and $G^2 = 7.92$ are also based on df = 3, and both have P-values just below 0.05. Hence there is some evidence of real differences among the operations in terms of relative frequencies of dumping, but not in terms of relative frequencies of slight and of moderate dumping among those cases where dumping does occur. Complement 2.4 describes another informative partition for Table 2.4.

In order for the chi-squared components to be independent, certain rules must be followed in choosing the component tables. For example, for an $r \times c$ table, one can compare rows 1 and 2, then combine rows 1 and 2 and compare them to row 3, then combine rows 1 through 3 and compare them to row 4, and so forth. In the last $2 \times c$ table formed, rows 1 through $r - 1$ combined are compared to row r. Each of the $r - 1$ comparisons has df $= c - 1$.

This partition can be generalized to one of $(r - 1)(c - 1)$ independent chi-squared statistics, each having df = 1. For instance, the G^2 components for the $(r - 1)(c - 1)$ 2×2 tables of the form

$$\begin{array}{c|c} n_{ij} & \sum_{b>j} n_{ib} \\ \hline \sum_{a>i} n_{aj} & \sum_{a>i} \sum_{b>j} n_{ab} \end{array}$$

sum exactly to G^2 for the entire $r \times c$ table. This partition is especially natural when both variables are ordinal. Other useful partitions for that case split X^2 or G^2 into two components, one that tests for a linear trend in the data and the other that tests for independence given that a linear trend model is appropriate. Partitions of this type are discussed in Chapters 5 and 7 and Complements 10.2 and 10.3.

Partitioning Models

Generally, G^2 components sum exactly to the G^2 value for the entire table, whereas the X^2 components need not sum to the X^2 value for the entire table. This additivity feature of the likelihood-ratio statistic G^2 will be used in the chapters on model building for decomposing the statistic into parts representing various effects. We will outline here the general method of partitioning the G^2 statistic in that context.

Suppose that there are two sets of ML estimates of expected frequencies, $\{\hat{m}_{ij}^{(1)}\}$ corresponding to model 1 and $\{\hat{m}_{ij}^{(2)}\}$ corresponding to model 2. Represent the likelihood-ratio statistics for testing goodness of fit of the models by $G^2[(1)]$ and $G^2[(2)]$. Suppose that model 2 is nested in model 1, in the sense that model 2 is a special case of model 1. In other words, model 2 is a simpler model than model 1. Then it is necessarily true (see Complement 2.5) that $G^2[(1)] \leq G^2[(2)]$. Now, $G^2[(2)] - G^2[(1)]$ is a measure of the distance of the best fit of model 2 from the best fit of model 1, and it gives an indication of how much better model 1 fits than model 2. We denote this difference by $G^2[(2)\,|\,(1)]$. Thus we have the representation

$$G^2[(2)] = G^2[(2)\,|\,(1)] + G^2[(1)] \tag{2.6}$$

Assume now that model 1 truly holds in the population. Then $G^2[(1)]$ has an asymptotic chi-squared distribution with $\text{df} = \text{df}_1$, say. If the simpler model 2 also holds, then $G^2[(2)]$ has an asymptotic chi-squared distribution with $\text{df} = \text{df}_2$, say. In this case Bishop et al. (1975, p. 525) show that $G^2[(2)\,|\,(1)] = G^2[(2)] - G^2[(1)]$ has an asymptotic chi-squared distribution with $\text{df} = \text{df}_2 - \text{df}_1$.

This result is partly of interest because the statistic $G^2[(2)\,|\,(1)]$ can be used to test the adequacy of the simpler model 2 under the assumption that model 1 holds. We will use this fact to test basic hypotheses such as independence for ordinal variables under the assumption that a certain structure (model 1 in this context) that reflects the ordinal nature of the data is appropriate. The decomposition is also of interest because Goodman (1970) showed that $G^2[(2)] - G^2[(1)]$ equals $2 \sum \sum \hat{m}_{ij}^{(1)} \log (\hat{m}_{ij}^{(1)}/\hat{m}_{ij}^{(2)})$ for the types of models we will study. Hence $G^2[(2)\,|\,(1)]$ has the same form as the likelihood-ratio statistic itself, with $\{\hat{m}_{ij}^{(1)}\}$ playing the role of the observed data. On the other hand, the difference of Pearson chi-squared statistics $X^2[(2)] - X^2[(1)]$ does not have the same Pearson form. In fact $X^2[(2)]$ is not even necessarily larger than $X^2[(1)]$ when model 2 is a special case of model 1.

2.4. ODDS RATIOS

Tests of significance are rarely adequate for answering all the questions we have about a data set. Researchers usually want to know more than whether two variables are independent. Indeed, in nearly all applications we would expect *some* association between two variables, even if it is very weak and substantively unimportant. Of greater practical interest are questions such as *how strong* is the association between family income and quality of mental health? Or, *how much* of a difference is there between the treatment and control groups in the success rate for treating a disease?

This section introduces a measure that describes the degree of association in a 2 × 2 table. It also is very useful for describing properties of models considered in later chapters. Refer to the general 2 × 2 population cross-classification in Table 2.2. Within row 1 the *odds* that the response is in column 2 instead of column 1 is defined to be

$$\Omega_1 = \frac{\pi_{12}}{\pi_{11}}$$

Within row 2 the corresponding odds equals

$$\Omega_2 = \frac{\pi_{22}}{\pi_{21}}$$

Each Ω_i is nonnegative, with value greater than 1.0 if response 2 is more likely than response 1. The ratio of these odds,

$$\theta = \frac{\Omega_2}{\Omega_1} = \frac{(\pi_{22}/\pi_{21})}{(\pi_{12}/\pi_{11})} = \frac{(\pi_{11}\pi_{22})}{(\pi_{12}\pi_{21})} \tag{2.7}$$

is referred to as the *odds ratio.* An alternative name for it is the *cross-product ratio*, since θ equals the ratio of the products $\pi_{11}\pi_{22}$ and $\pi_{12}\pi_{21}$ of proportions from cells that are diagonally opposite.

Properties

Note that each odds Ω_i can be expressed as

$$\Omega_i = \frac{\pi_{2(i)}}{\pi_{1(i)}}$$

and thus

$$\theta = \frac{(\pi_{2(2)}/\pi_{1(2)})}{(\pi_{2(1)}/\pi_{1(1)})} \tag{2.8}$$

The two conditional distributions $(\pi_{1(1)}, \pi_{2(1)})$ and $(\pi_{1(2)}, \pi_{2(2)})$ are identical, and hence the variables are independent, if and only if $\Omega_1 = \Omega_2$. In this case the odds ratio $\theta = 1.0$. If $1 < \theta < \infty$, individuals in row 2 are more likely to make the second response than are individuals in row 1; that is, $\pi_{2(2)} > \pi_{2(1)}$. If $0 \leq \theta < 1$, individuals in row 2 are less likely to make the second response than are individuals in row 1; that is, $\pi_{2(2)} < \pi_{2(1)}$.

In most applications the population proportions $\{\pi_{ij}\}$ are unknown parameters, and hence so is θ. For sample cell frequencies $\{n_{ij}\}$, a sample analog of θ is

$$\hat{\theta} = \frac{n_{11}n_{22}}{n_{12}n_{21}} \tag{2.9}$$

An interesting property of $\hat{\theta}$ is that its value does not change if both cell frequencies within any row are multiplied by a nonzero constant, or if both cell frequencies within any column are multiplied by a constant. Hence $\hat{\theta}$ estimates the same characteristic (θ) even if disproportionately large or small samples are selected from the various marginal categories of a variable. In particular, it estimates the same characteristic regardless of whether sampling is full multinomial or independent multinomial. The measure also takes the same value if the orientation of the table is reversed so that the rows become the columns and the columns become the rows. Edwards (1963) showed that the only measures for 2×2 tables that have these properties are $\hat{\theta}$ and functions of it.

If the order of the rows is reversed or if the order of the columns is reversed, the new value of $\hat{\theta}$ is simply the inverse of the original value. Hence two possible values for $\hat{\theta}$ represent the same degree of association (but in opposite directions) if one value is the inverse of the other.

From (2.9), $\hat{\theta}$ equals 0 or ∞ if any of the $n_{ij} = 0$. These are undesirable estimates of θ if there is no theoretical reason to expect any of the π_{ij} to equal zero. If the subjects in each row were randomly selected, Gart and Zweiful (1967) showed that the estimator

$$\tilde{\theta} = \frac{(n_{11} + 0.5)(n_{22} + 0.5)}{(n_{12} + 0.5)(n_{21} + 0.5)} \tag{2.10}$$

has smaller bias and smaller mean square error.

The odds ratio is a multiplicative function of the cell proportions. Its logarithm is an additive function, namely,

$$\log \theta = \log \pi_{11} - \log \pi_{12} - \log \pi_{21} + \log \pi_{22} \tag{2.11}$$

and it may equal any real number. Here, and in the rest of the text, log denotes the natural logarithm. The log odds ratio is symmetric around the independence value of 0.0 in the sense that a reversal of the two rows or of the two columns results in a change in its sign. Hence two values for $\log \theta$ that are the same except for sign (such as -4 and 4) represent the same degree of association.

Under the standard random sampling models both $\hat{\theta}$ and $\log \hat{\theta}$ are asymptotically (as $n \to \infty$) normally distributed around their population values. However, $\log \hat{\theta}$, being an additive rather than multiplicative function of the $\{n_{ij}\}$, tends to converge faster than does $\hat{\theta}$ to its asymptotic distribution. In particular, the distribution of $\hat{\theta}$ is highly skewed for small n. If $\theta \leq 1$, for instance, $\hat{\theta}$ cannot be much smaller than θ (since $\hat{\theta} \geq 0$), but it could be much larger than θ with nonnegligible probability.

The asymptotic standard deviation of $\log \hat{\theta}$, denoted by $\sigma(\log \hat{\theta})$, can be estimated by

$$\hat{\sigma}(\log \hat{\theta}) = \sqrt{\frac{1}{n_{11}} + \frac{1}{n_{12}} + \frac{1}{n_{21}} + \frac{1}{n_{22}}} \tag{2.12}$$

An approximate $100(1-p)$ percent confidence interval for $\log \theta$ is given by

$$\log \hat{\theta} \pm z_{p/2}\, \hat{\sigma}(\log \hat{\theta}) \tag{2.13}$$

where $z_{p/2}$ is the percentage point from the standard normal distribution corresponding to a two-tail probability equal to p. The corresponding confidence interval for θ can be obtained by exponentiating endpoints of the confidence interval for $\log \theta$. One should not form confidence intervals for θ directly using $\hat{\theta}$ and its standard error (1) because of its slower convergence to normality and (2) because this confidence interval is not equivalent to the one obtained using $1/\hat{\theta}$ and its standard error; that is, the values in one interval are not identically the inverses of the values in the other. Again the estimates of θ and of $\sigma(\log \hat{\theta})$ have smaller asymptotic bias and mean square error if the $\{n_{ij}\}$ are replaced by $\{n_{ij} + 0.5\}$. See Fleiss (1979) for a discussion of alternative ways of obtaining a confidence interval for θ.

Death Penalty Example

For the death penalty data (Table 2.1) introduced in Section 2.1, the sample odds ratio equals $\hat{\theta} = (19 \times 149)/(141 \times 17) = 1.18$. This value means that the odds of getting the death penalty were 1.18 times higher for the white defendants in the sample than for the black defendants. The estimated standard error of $\log \hat{\theta} = 0.166$ is

$$\hat{\sigma}(\log \hat{\theta}) = \sqrt{\frac{1}{19} + \frac{1}{141} + \frac{1}{17} + \frac{1}{149}} = 0.354$$

An approximate 95 percent confidence interval for the value of $\log \theta$ in the hypothetical population from which we view this sample as having been randomly selected is $0.166 \pm 1.96(0.354)$, or $(-0.527, 0.860)$. The corresponding confidence interval for θ is $(e^{-0.527}, e^{0.860})$, or $(0.59, 2.36)$. Since 1.0 is in the confidence interval for θ (and 0.0 is in the confidence interval for $\log \theta$), it is plausible that death penalty verdict is independent of defendant's race. The data do not imply that white defendants and black defendants have different likelihoods of receiving the death penalty.

When 0.5 is added to each cell, the estimate of $\log \theta$ becomes 0.163 with a standard error of 0.350. The 95 percent confidence interval for θ becomes (0.59, 2.34). This interval is very similar to the one previously obtained, since none of the cell frequencies is especially small.

Odds Ratios for $r \times c$ Tables

The discussion of odds ratios has so far focused on 2×2 tables. For the general $r \times c$ table odds ratios can be formed using each of the $\binom{r}{2} = r(r-1)/2$ pairs of rows in combination with each of the $\binom{c}{2} = c(c-1)/2$ pairs of columns. For rows a and b and columns c and d the odds ratio $(\pi_{ac}\pi_{bd})/(\pi_{bc}\pi_{ad})$ uses four cells occurring in a rectangular pattern, and there are $\binom{r}{2}\binom{c}{2}$ odds ratios of this type. The independence of the two variables is equivalent to the condition that all these population odds ratios equal 1.

There is much redundancy of information when the entire set of these odds ratios is used to characterize the association in a table. For example, consider the set of $(r-1)(c-1)$ odds ratios

$$\frac{(\pi_{ij}\pi_{rc})}{(\pi_{rj}\pi_{ic})}, \qquad i = 1, \ldots, r-1, \qquad j = 1, \ldots, c-1 \tag{2.14}$$

Each odds ratio is formed using the rectangular array of cells determined by rows i and r and columns j and c, as indicated in Figure 2.1. It can be shown that the $(r-1)(c-1)$ odds ratios in (2.14) determine all $\binom{r}{2}\binom{c}{2}$ odds ratios that can be formed from pairs of rows and pairs of columns. Independence of the two variables is therefore also equivalent to the condition that the odds ratios in the basic set (2.14) are identically equal to one.

Odds Ratios for Ordinal Variables

The construction (2.14) for forming a minimal set of odds ratios that determine the entire set is not unique. Another basic set of $(r-1)(c-1)$ odds ratios, usually more natural for cross-classifications of ordinal variables, is

$$\theta_{ij} = \frac{(\pi_{ij}\pi_{i+1,\,j+1})}{(\pi_{i,\,j+1}\pi_{i+1,\,j})}, \qquad i = 1, \ldots, r-1, \qquad j = 1, \ldots, c-1 \tag{2.15}$$

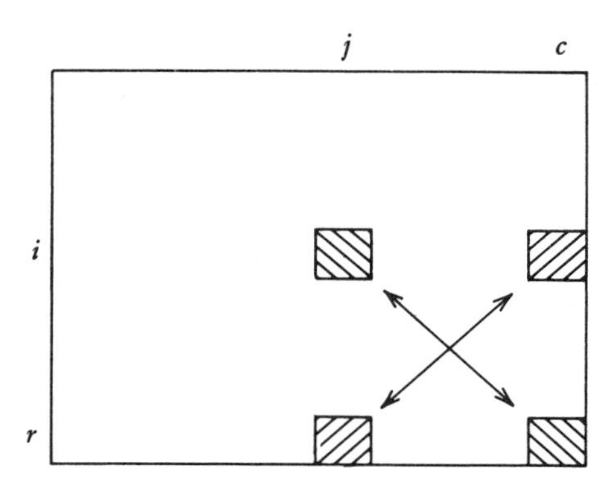

Figure 2.1. Odds ratios (2.14).

These odds ratios are formed using cells in adjacent rows and adjacent columns. Their values describe the relative magnitudes of "local" associations in the table. Models for ordinal variables that have been formulated in terms of the $\{\theta_{ij}\}$ are discussed in Chapter 5.

The local odds ratios treat row and column variables alike. Another family of odds ratios, one that makes a distinction between rows and columns, is

$$\theta'_{ij} = \frac{\left(\sum_{b \leq j} \pi_{ib}\right)\left(\sum_{b > j} \pi_{i+1,b}\right)}{\left(\sum_{b > j} \pi_{ib}\right)\left(\sum_{b \leq j} \pi_{i+1,b}\right)}$$

$$= \frac{[F_{j(i)}/(1 - F_{j(i)})]}{[F_{j(i+1)}/(1 - F_{j(i+1)})]},$$

$$i = 1, \ldots, r-1, \qquad j = 1, \ldots, c-1 \qquad (2.16)$$

These odds ratios are local in the row variable but "global" in the column variable, since all c categories of the column variable are used in each odds ratio. For an adjacent pair of rows i and $i + 1$,

$$\log \theta'_{ij} \geq 0 \quad (\text{or} \quad \theta'_{ij} \geq 1), \qquad \text{for} \qquad j = 1, \ldots, c-1$$

is equivalent to

$$F_{j(i)} \geq F_{j(i+1)}, \qquad \text{for} \qquad j = 1, \ldots, c \qquad (2.17)$$

When (2.17) holds, the distribution in row $i + 1$ is said to be *stochastically higher* than the one in row i. This means that relatively more of the probability in row $i + 1$ falls at the high end of the ordinal scale for the column variable. Figure 2.2 illustrates stochastic orderings for underlying continu-

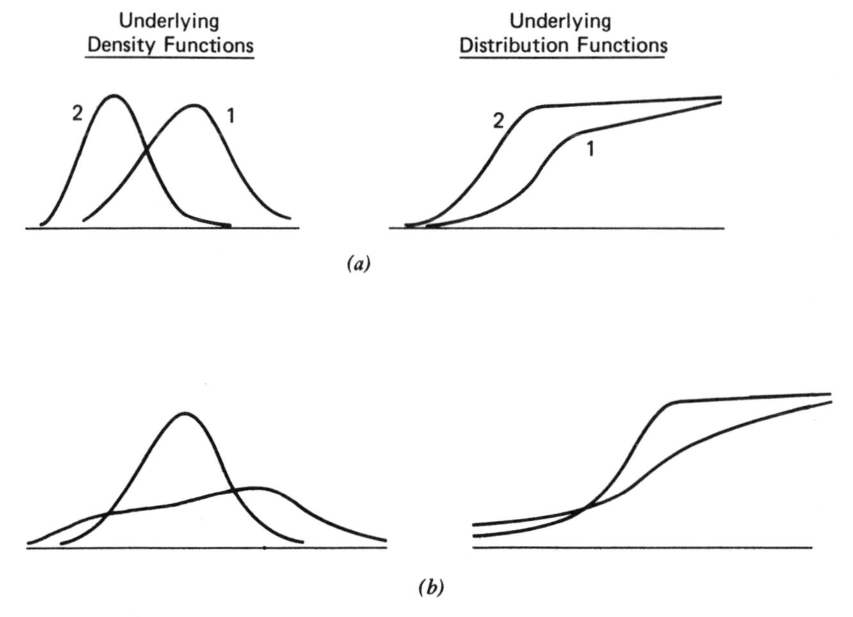

Figure 2.2. (*a*) Distribution 1 stochastically higher than distribution 2. (*b*) Distributions not stochastically ordered.

ous density functions corresponding to the $\{\pi_{j(i)}\}$ and for underlying continuous distribution functions corresponding to the $\{F_{j(i)}\}$.

The $\{\theta'_{ij}\}$ are particularly meaningful when a distinction is made between response and explanatory variables. For example, they can be used to compare pairs of rows with respect to their distribution on an ordinal response variable. Their values indicate whether the rows are stochastically ordered on that response. Models for the $\{\theta'_{ij}\}$ that treat the column variable as a response variable are considered in Chapter 7.

A third family of odds ratios for cross-classifications of ordinal variables is

$$\theta''_{ij} = \frac{\left(\sum_{a \leqslant i} \sum_{b \leqslant j} \pi_{ab}\right)\left(\sum_{a > i} \sum_{b > j} \pi_{ab}\right)}{\left(\sum_{a \leqslant i} \sum_{b > j} \pi_{ab}\right)\left(\sum_{a > i} \sum_{b \leqslant j} \pi_{ab}\right)}$$

$$= \frac{F_{ij}(1 + F_{ij} - F_{i+} - F_{+j})}{(F_{i+} - F_{ij})(F_{+j} - F_{ij})},$$

$$i = 1, \ldots, r-1, \qquad j = 1, \ldots, c-1 \tag{2.18}$$

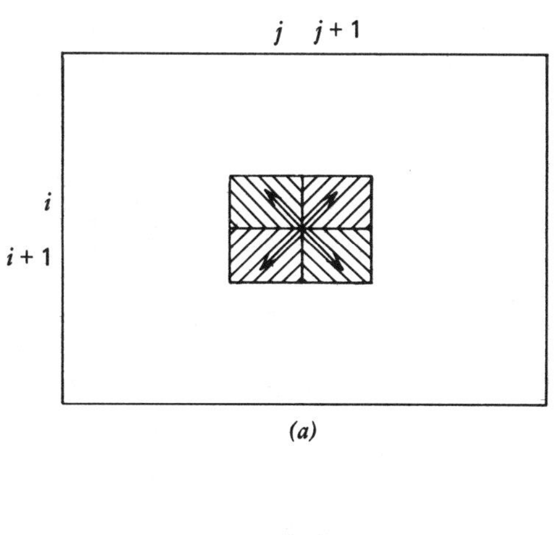

(a)

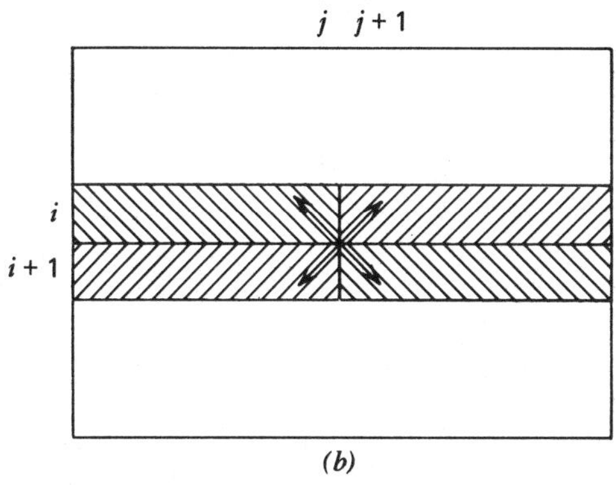

(b)

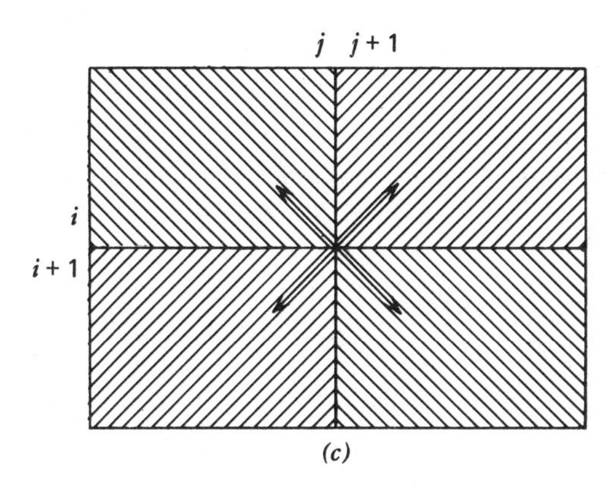

(c)

Figure 2.3. Three sets of $(r-1)(c-1)$ odds ratios for ordinal variables: (a) θ_{ij}; (b) θ'_{ij}; (c) θ''_{ij}.

These measures are the regular odds ratios computed for the 2×2 tables corresponding to the $(r-1)(c-1)$ ways of collapsing the row and column classifications into dichotomies. They treat row and column variables alike and describe associations that are global in both variables. Section 8.1 describes models for these odds ratios.

Figure 2.3 illustrates the local, local-global, and global odds ratios. For each set, independence is equivalent to all log odds ratios equaling zero. An association described by one of these measures is referred to as "positive" or "negative" in accordance with the sign of the log odds ratio. If all $\log \theta_{ij} \geq 0$, then all $\log \theta'_{ij} \geq 0$. If all $\log \theta'_{ij} \geq 0$, then all $\log \theta''_{ij} \geq 0$. The converses of these statements are not true. The condition that all local log odds ratios be positive is therefore the most stringent of three possible definitions for "uniformly positive association."

The less localized the odds ratio, the more precise its sample value tends to be as an estimator of its population value, since formula (2.12) is applied to larger sample totals. Hence, if the $\{\theta_{ij}\}$ are approximately equal, if the $\{\theta'_{ij}\}$ are approximately equal, and if the $\{\theta''_{ij}\}$ are approximately equal, the sample estimates of the third set will tend to be smoothest.

Conversion of the cell probabilities into the set of odds ratios (2.14), (2.15), (2.16), or (2.18) does not result in a loss of information. Given the marginal probabilities $\{\pi_{i+}\}$ and $\{\pi_{+j}\}$, the cell probabilities are determined by these odds ratios (see Plackett 1981, p. 35, and Complement 4.5).

Dumping Severity Data

We will illustrate these three types of odds ratios for ordinal variables using the dumping severity data that were introduced in the previous section (see Table 2.4). Table 2.5 contains the sample values $\{\hat{\theta}_{ij}\}$, $\{\hat{\theta}'_{ij}\}$, and $\{\hat{\theta}''_{ij}\}$ of the ordinal odds ratios. The values $\hat{\theta}_{12}$, $\hat{\theta}'_{12}$, and $\hat{\theta}''_{12}$ mean, respectively:

1. The estimated odds that dumping is moderate instead of slight is 2.26 times higher for operation B than for operation A.
2. The estimated odds that dumping is moderate instead of none or slight is 1.82 times higher for operation B than for operation A.
3. The estimated odds that dumping is moderate instead of none or slight is 1.86 times higher when some stomach is removed (operations B, C, D) than when none is removed (operation A).

All three sets of measures indicate a generally positive association, though the $\{\hat{\theta}''_{ij}\}$ show the most consistency. From the $\{\hat{\theta}'_{ij}\}$ in this sample, operation D is stochastically higher on the dumping severity scale than operation C. Similar comparisons of nonadjacent rows reveal that for this

Table 2.5. Values of Ordinal Odds Ratios for Dumping Severity Data

	$\hat{\theta}_{ij}$		$\hat{\theta}'_{ij}$		$\hat{\theta}''_{ij}$	
i \ j	1	2	1	2	1	2
1	0.74	2.26	0.92	1.82	1.38	1.86
2	2.04	0.53	1.69	0.86	1.74	1.33
3	1.04	1.40	1.14	1.44	1.55	1.53

sample, operation D is stochastically higher than operations A and B, and operation C is stochastically higher than operation A.

2.5. MEASURING ASSOCIATION

Section 2.4 introduced the odds ratio measure of association. This section briefly discusses other ways of describing departures from independence in a two-way table. Chapters 9 and 10 consider this matter in much greater detail for the case where at least one of the variables is ordinal.

Norms of Chi-Squared

If the variables are dependent, the asymptotic expectations of the X^2 and G^2 statistics are proportional to the sample size n. Hence these statistics cannot by themselves be used to measure *strength* of association, since even a trivial departure from independence results in an impressively large chi-squared statistic if the sample size is large enough. One remedy is to adjust the statistic by dividing it by some multiple of the sample size.

Cramér (1946, p. 282) showed that the X^2 statistic cannot exceed the sample size multiplied by the minimum of $r-1$ and $c-1$. This bound suggests the measure of association

$$V^2 = \frac{X^2}{[n \min(r-1, c-1)]} \tag{2.19}$$

This measure, called *Cramér's* V^2, falls between 0 and 1 with larger values representing stronger association. For the 2×2 table, $V^2 = X^2/n$ is also called ϕ^2, and it is identical to the square of the Pearson correlation coefficient obtained by assigning numeric scores to the rows and to the columns.

Cramér's V^2 is a simple measure. It cannot be expressed simply in terms of probabilities or odds, however, so it is difficult to interpret its absolute magnitude. It is not obvious how to interpret the value $V^2 = 10.54/(417 \times 2) = 0.013$ for the dumping severity data of Table 2.4. Hence Cramér's V^2 is mainly useful for *comparing* strengths of association in tables of the same dimensions. In other words, it is more useful for describing relative strengths of association than absolute strengths. If two tables have differing sample sizes, their X^2 values are not comparable, but their V^2 values are. Even for this purpose, though, it is hazardous to use V^2 if the marginal distributions in one table are very different from the marginal distributions in another table.

Difference of Proportions

Other measures of association, not based on the chi-squared statistic, can be more easily interpreted than V^2. Many studies are mainly concerned with investigating whether two groups have differing proportions of responses in a certain category. A simple and easily interpretable measure for that case is the *difference of proportions.*

In Table 2.1 the proportion of white defendants who received the death penalty was $p_{1(1)} = 0.119$, and the proportion of black defendants who received the death penalty was $p_{1(2)} = 0.102$. The difference of proportions $p_{1(1)} - p_{1(2)} = 0.017$ describes the weak association between defendant's race and death penalty verdict in that sample. The estimated standard error

$$\hat{\sigma}(p_{1(1)} - p_{1(2)}) = \sqrt{\frac{p_{1(1)}(1 - p_{1(1)})}{n_{1+}} + \frac{p_{1(2)}(1 - p_{1(2)})}{n_{2+}}}$$

equals 0.035, so a true difference $\pi_{1(1)} - \pi_{1(2)}$ of zero is well within the limits of sampling error.

A difference in proportions of a certain size sometimes has greater substantive importance when $\pi_{1(1)}$ and $\pi_{1(2)}$ are both very close to 0 or 1 than if they are not. For instance, in a comparison of two drugs on the proportion of individuals who suffer bad side effects with their use, the difference between 0.06 and 0.03 may seem more important than the difference between 0.50 and 0.47. In such cases the odds ratio may better reflect the association.

Proportional Reduction in Error

Goodman and Kruskal (1954, 1959, 1963, 1972) have introduced measures of association that have proportional reduction in error interpretations similar to the one given the coefficient of determination for interval variables. Two of these measures, referred to as lambda and tau, are designed for cross-classifications of nominal variables. The measure gamma, which will be discussed in Chapters 9 and 10, is appropriate for ordinal variables. We feel that these measures are superior to measures based on chi-squared. For an introduction to the proportional reduction in error approach for constructing measures of association, see Goodman and Kruskal (1954) or Chapter 8 of the text by Agresti and Agresti (1979).

2.6. LOGLINEAR MODELS FOR TWO DIMENSIONS

From Section 2.1, two variables are independent if $\pi_{ij} = \pi_{i+}\pi_{+j}$ for all i and j. The corresponding expression for the expected frequencies $\{m_{ij} = n\pi_{ij}\}$ is $m_{ij} = n\pi_{i+}\pi_{+j}$ for all i and j. On a logarithmic scale independence is equivalent to the additive relationship

$$\log m_{ij} = \log n + \log \pi_{i+} + \log \pi_{+j} \tag{2.20}$$

If two variables are independent therefore, the log expected frequency for cell (i, j) is an additive function of an ith row effect and a jth column effect. An alternative formulation of model (2.20) is

$$\log m_{ij} = \mu + \lambda_i^X + \lambda_j^Y \tag{2.21}$$

This model is called the *loglinear model* for independence in a two-way table.

We now consider a more general loglinear model for two variables, one that corresponds to dependence. Let $\mu_{ij} = \log m_{ij}$, let

$$\mu_{i\cdot} = \sum_j \frac{\mu_{ij}}{c}, \qquad \mu_{\cdot j} = \sum_i \frac{\mu_{ij}}{r}$$

and let

$$\mu = \mu_{\cdot\cdot} = \sum_i \sum_j \frac{\mu_{ij}}{(rc)}$$

denote the overall mean of the $\{\log m_{ij}\}$. Then if

$$\begin{aligned} \lambda_i^X &= \mu_{i\cdot} - \mu \\ \lambda_j^Y &= \mu_{\cdot j} - \mu \\ \lambda_{ij}^{XY} &= \mu_{ij} - \mu_{i\cdot} - \mu_{\cdot j} + \mu \end{aligned} \tag{2.22}$$

it follows from simple algebra that

$$\log m_{ij} = \mu + \lambda_i^X + \lambda_j^Y + \lambda_{ij}^{XY} \tag{2.23}$$

This is the most general model for two dimensions, since a model of this form provides a perfect fit for any set of positive expected frequencies $\{m_{ij}\}$.

The notation used in model (2.23) is similar to that for the two-way analysis of variance model. The $\{\lambda_i^X\}$ and $\{\lambda_j^Y\}$ are scaled so that they are deviations about a mean, and hence

$$\sum_i \lambda_i^X = \sum_j \lambda_j^Y = 0$$

Thus there are $r - 1$ linearly independent row effect parameters and $c - 1$ linearly independent column effect parameters. The $\{\lambda_i^X\}$ and $\{\lambda_j^Y\}$ pertain to the relative numbers of cases in the categories of X and Y. If $\lambda_i^X > 0$, for instance, the average of the expected frequencies in the ith row is larger than the average of the expected frequencies over the entire table. This average is computed on the log scale, since λ_i^X refers to the relative size of the arithmetic mean of the $\{\log m_{ij},\ j = 1, \ldots, c\}$. In the independence model note that for a $2 \times c$ table

$$\log m_{1j} - \log m_{2j} = \lambda_1^X - \lambda_2^X = 2\lambda_1^X$$

since $\lambda_1^X = -\lambda_2^X$. Hence $\log (m_{1j}/m_{2j})$ is the same in every column when the variables are independent, and $\exp (2\lambda_1^X)$ can be interpreted as the common value of the odds.

The $\{\lambda_{ij}^{XY}\}$ satisfy

$$\sum_i \lambda_{ij}^{XY} = \sum_j \lambda_{ij}^{XY} = 0$$

so that $(r - 1)(c - 1)$ of these terms are linearly independent. The independence model is the special case of the general model (2.23) in which all $\lambda_{ij}^{XY} = 0$. Hence the $\{\lambda_{ij}^{XY}\}$ are "association parameters" that reflect departures from independence.

The independence model, $\log m_{ij} = \mu + \lambda_i^X + \lambda_j^Y$, has $1 + (r - 1) + (c - 1) = r + c - 1$ linearly independent parameters. The general model has the additional $(r - 1)(c - 1)$ $\{\lambda_{ij}^{XY}\}$ association parameters. The total number of parameters in the general model equals $1 + (r - 1) + (c - 1) + (r - 1)(c - 1) = rc$, which is the total number of cells in the table. For tables of any number of dimensions, the general loglinear model has as many parameters as there are cells in the table and is called a *saturated* model.

There are direct relationships between parameters in loglinear models and the odds ratio measure θ. This relationship is simplest for the 2×2 table. There the constraints on the association parameters imply that

$$\lambda_{11}^{XY} = \lambda_{22}^{XY} = -\lambda_{12}^{XY} = -\lambda_{21}^{XY}$$

and from their definition (2.22) it follows that

$$\lambda_{11}^{XY} = \left(\frac{1}{4}\right) \log \left[\frac{(m_{11} m_{22})}{(m_{12} m_{21})}\right] = \left(\frac{1}{4}\right) \log \theta \tag{2.24}$$

Thus the log odds ratio for a 2×2 table equals four times the association parameter in the saturated loglinear model. If $\lambda_{ij}^{XY} > 0$ the expected frequency in cell (i, j) is greater than the value $n\pi_{i+}\pi_{+j}$ corresponding to independence.

NOTES

Section 2.2

1. Pearson mistakenly suggested $\text{df} = rc - 1$ for the chi-squared test of independence and was corrected by R. A. Fisher in 1922. See Joan Fisher Box (1978, pp. 84–88) for an interesting discussion of the controversy surrounding the correct formula for df.
2. The asymptotic null equivalence of X^2 and G^2 means that $n(G^2 - X^2)$ converges in probability to zero as $n \to \infty$. See Bishop et al. (1975, p. 514). Both tests have the same asymptotic power properties for contiguous alternatives, and hence they are equally as efficient in the Pitman sense. See Haberman (1974a, p. 101). They do not have the same asymptotic power for fixed alternatives, since their noncentrality parameters differ.
3. For 2×2 tables, X^2 can be expressed as $X^2 = n(n_{11}n_{22} - n_{12}n_{21})^2/(n_{1+}n_{2+}n_{+1}n_{+2})$.
4. Given the marginal distributions of a 2×2 table, if the variables are independent the distribution of n_{11} has the hypergeometric form

$$P(n_{11} \mid n_{1+}, n_{2+}, n_{+1}, n_{+2}) = \frac{\binom{n_{1+}}{n_{11}}\binom{n_{2+}}{n_{+1} - n_{11}}}{\binom{n}{n_{+1}}}$$

where n_{11} takes on values between $m = \max(0, n_{1+} + n_{+1} - n)$ and $M = \min(n_{1+}, n_{+1})$. Thus, to test H_0: $\theta = 1$ (independence) against H_a: $\theta > 1$, one can use as an attained significance level,

$$P = \sum_{i=n_{11}}^{M} P(i \mid n_{1+}, n_{2+}, n_{+1}, n_{+2})$$

This conditional test of independence is referred to as Fisher's exact test (see Fisher 1934). Agresti and Wackerly (1977) considered generalizations of this test for $r \times c$ tables.

5. For 2 × 2 tables Yates (1934) suggested a correction to the Pearson statistic,

$$X_c^2 = \sum_i \sum_j \frac{(|n_{ij} - \hat{m}_{ij}| - 0.5)^2}{\hat{m}_{ij}}$$

to adjust for using the continuous chi-squared distribution to approximate a discrete distribution. Recent studies have indicated that the uncorrected statistic has more nearly a chi-squared distribution than the corrected one (e.g., see Conover, 1974). The corrected statistic seems to be mainly useful for giving P-values (from the chi-squared distribution) that are more similar to the hypergeometric probabilities obtained with Fisher's exact test.

Section 2.3

6. Independent chi-squared statistics are *not* obtained when the component tables in a partitioning compare each level of a variable to a particular fixed level. However, if this type of comparison is natural, one can use a Bonferroni approach to control the overall type I error probability in testing independence for each component table. See Brunden (1972).

Section 2.4

7. The invariance properties of $\hat{\theta}$ imply that it makes sense as an estimate of θ regardless of whether a study is retrospective, prospective, or cross-sectional in nature. See Fienberg (1980, p. 135), Fleiss (1981, p. 27), or Plackett (1981, pp. 26–27).
8. For 2 × 2 tables a test of independence can also be based on the odds ratio. The asymptotic normality of log $\hat{\theta}$ implies that $z = (\log \hat{\theta})/\hat{\sigma}(\log \hat{\theta})$ has an approximate standard normal distribution when H_0 is true. Thus z^2 is an approximate chi-squared statistic based on df = 1.
9. Another commonly used measure for 2 × 2 tables is the ratio of probabilities $\pi_{1(1)}/\pi_{1(2)}$, called the *relative risk*. When both probabilities are close to zero, this measure is similar in value to the odds ratio.
10. Lehmann (1966) defined two random variables (X, Y) to be *positively quadrant dependent* if

$$P(X \le x, Y \le y) \ge P(X \le x)P(Y \le y) \quad \text{all } x \text{ and } y$$

and *positively likelihood ratio dependent* if their joint density f satisfies

$$f(x_1, y_1)f(x_2, y_2) \geq f(x_1, y_2)f(x_2, y_1) \quad \text{whenever } x_1 < x_2 \text{ and } y_1 < y_2$$

He defined Y to be *positively regression dependent* on X if

$$P(Y \leq y \mid X = x) \quad \text{is nonincreasing in } x$$

For cross-classifications of ordinal variables, positive quadrant dependence corresponds to nonnegative values for all global log odds ratios $\{\log \theta''_{ij}\}$, positive likelihood ratio dependence corresponds to nonnegative values for all local log odds ratios $\{\log \theta_{ij}\}$, and positive regression dependence corresponds to nonnegative values for all local-global log odds ratios $\{\log \theta'_{ij}\}$.

Section 2.5

11. To reduce the dependence of measures of association on the marginal distributions, one can calculate their values for an adjusted table having uniform marginal frequencies. See Mosteller (1968), Smith (1976), and Agresti (1981b).

Section 2.6

12. If $\lambda_i^X > 0$, the geometric mean $(m_{i1} \cdots m_{ic})^{1/c}$ in the ith row is larger than the geometric mean $(m_{11} \cdots m_{rc})^{1/rc}$ computed over the entire table.

COMPLEMENTS

1. Let $\alpha_{ij} = \pi_{ij}\pi_{rc}/\pi_{ic}\pi_{rj}$, $1 \leq i \leq r-1$, $1 \leq j \leq c-1$.

a. Show that the $\{\alpha_{ij}\}$ determine all $\binom{r}{2}\binom{c}{2}$ odds ratios that can be formed from pairs of rows and pairs of columns.

b. Show that the $\{\alpha_{ij}\}$ determine the local odds ratios, and vice versa.

2. For the local odds ratios, show that $\theta_{ij} \geq 1$, $1 \leq j \leq c-1$, implies that the conditional distribution in row $i+1$ is stochastically higher than the conditional distribution in row i. Show that the converse is not true.

3. Define a family of odds ratios corresponding to the partitioning described in Section 2.3 that produces $(r-1)(c-1)$ independent chi-squared statistics. Give some properties of these odds ratios, and give an example of a $2 \times c$ table for which they would be of greater interest than the odds ratios discussed in Section 2.4.

4. For Table 2.4, partition chi-squared into three parts by comparing operation A to operation B, operation C to operation D, and operations A and B combined to operations C and D combined. What is your interpretation?

5. For the full multinomial sampling model,

a. Show that the kernel of the log likelihood function (i.e., the part involving the parameters $\{\pi_{ij} = m_{ij}/n\}$) can be expressed as $L(\mathbf{m}) = \sum\sum n_{ij} \log m_{ij}$.

b. Show that if model 2 is nested in model 1, then $L(\hat{\mathbf{m}}^{(1)}) \geq L(\hat{\mathbf{m}}^{(2)})$, and therefore $G^2[(1)] \leq G^2[(2)]$. Note that $G^2[(2)\,|\,(1)] = -[L(\hat{\mathbf{m}}^{(2)}) - L(\hat{\mathbf{m}}^{(1)})]$.

6. For two conditional distributions $\{\pi_{j(i)}, j = 1, \ldots, c-1, i = 1, 2\}$ with $F_{j(i)} = \sum_{b \leqslant j} \pi_{b(i)}$, the plot of lines that connects successively the points $\{(0, 0), (F_{1(1)}, F_{1(2)}), (F_{2(1)}, F_{2(2)}), (F_{3(1)}, F_{3(2)}), \ldots, (1.0, 1.0)\}$ is called a *cumulative sum diagram* (CSD). See Grove (1980).

a. Show that a straight line for the CSD corresponds to independence.

b. Show that a convex CSD corresponds to the condition that all local log odds ratios for the two rows are nonnegative.

c. Draw and interpret the sample CSD for the following data (Source: Armitage 1955. Reprinted with permission from the Biometric Society.):

	Change in Size of Ulcer Crater				
Treatment Group	Larger	$<\frac{2}{3}$ Healed	$\geq\frac{2}{3}$ Healed	Healed	Total
A	12	10	4	6	32
B	5	8	8	11	32

CHAPTER 3

Associations in Multidimensional Tables

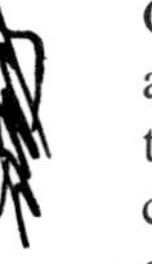

Chapter 2 presented elementary ways of testing for and measuring association in a two-way cross-classification table. In most studies it is important to study how *several* variables interrelate. Inadequate and even incorrect conclusions can result from studying the variables only two at a time. This chapter introduces definitions and concepts that are needed for the analysis of multidimensional cross-classification tables.

Section 3.1 shows that an association between two variables can change drastically when a third variable is controlled. Section 3.2 formally discusses conditions under which associations change and introduces the concept of three-factor interaction for categorical variables. Possible association patterns in three-dimensional tables are described in Section 3.3. That section introduces loglinear models that represent the association patterns. The following chapter shows how to use sample data to fit these models. Chapters 3 and 4 give the traditional development of this topic whereby all variables are treated as nominal. In Chapter 5 we shall describe some recent research that shows how these approaches can be modified for ordinal variables.

3.1. PARTIAL ASSOCIATION

In most applications there are several relevant variables, and a multidimensional table is used to display the data in their most complete form. When we study the relationship between a response variable Y and an explanatory variable X, for instance, there are usually other variables ("covariates") whose effects we want to control because of their possible influence on that relationship. In a comparison of survival times of patients who have undergone one of two treatments for heart disease, it would be important to control for characteristics such as age, sex, quality of health, and various aspects of the patient's medical history.

Cross sections of a multidimensional table display the distribution of $X-Y$ cell counts at various levels of another variable (or at the various combinations of levels of other variables). These cross sections are referred to as *partial tables*. In the partial tables the other variable(s) is controlled in the sense that its value is held constant. The two-dimensional table obtained by summing corresponding counts in the partial tables is called the *$X-Y$ marginal table*. That table, the type considered in Chapter 2, ignores rather than controls the other variables.

The association displayed in a partial table can be analyzed using the methods introduced in Chapter 2. However, the information conveyed by $X-Y$ partial tables can be quite different from that conveyed by the corresponding $X-Y$ marginal table. In fact it can be quite misleading to study only the two-dimensional marginal distributions of a multidimensional table.

To illustrate these points, we return to the death penalty data introduced in Section 2.1. We studied the bivariate relationship between the death penalty verdict and defendant's race using Table 2.1. That table is actually a two-dimensional marginal table of a three-dimensional cross-classification that Radelet presented in which the variable "victim's race" was also included. This expanded cross-classification is given in Table 3.1. The marginal cross-classification given in Table 2.1 is obtained from Table 3.1 by summing the frequencies over the two categories of victim's race (i.e., $19 + 0$, $132 + 9$, $11 + 6$, $52 + 97$).

For each of the four combinations of defendant's race and victim's race Table 3.1 lists the proportion of subjects who received the death penalty. For white defendants the death penalty was imposed over 12 percentage

Table 3.1. Death Penalty Verdict by Defendant's Race and Victim's Race

Defendant's Race	Victim's Race	Death Penalty Yes	Death Penalty No	Proportion Yes
White	White	19	132	0.126
	Black	0	9	0.000
Black	White	11	52	0.175
	Black	6	97	0.058

Source: Radelet (1981). Reprinted by permission.

points more often when the victim was white than when the victim was black. (In fact, the death penalty was never imposed in this sample when a white killed a black.) Similarly for black defendants the death penalty was imposed about 12 percentage points more often when the victim was white than when the victim was black. Controlling for defendant's race therefore, the death penalty verdict seems to be more likely when a white is killed than when a black is killed.

Now, consider the association between defendant's race and the death penalty verdict, controlling for victim's race. When the victim was white, the death penalty was imposed about 5 percentage points more often when the defendant was black than when the defendant was white. When the victim was black, the death penalty was imposed over 5 percentage points more often when the defendant was black than when the defendant was white. In summary, controlling for victim's race, black defendants were somewhat more likely to be given the death penalty than were white defendants. On the surface this conclusion seems contradictory to the observation made in Section 2.1 that the sample proportion receiving the death penalty was higher for white defendants than for black defendants.

How can we explain this very different association between death penalty verdict and defendant's race obtained when controlling for victim's race? Let us study Table 3.2, which lists for each pair of variables the marginal odds ratio and also the partial odds ratio at each of the two levels of the third variable. The marginal odds ratios describe the association when the third variable is ignored (i.e., when we sum the counts over the levels of the third variable to get a marginal two-way table). The partial odds ratios describe the association when the third variable is controlled. Since one of the cell counts in the three-dimensional table equals zero and since several of them are small, 0.5 was added to each cell count before these odds ratios were computed.

Table 3.2. Odds Ratios for Death Penalty (*P*), Victim's Race (*V*), and Defendant's Race (*D*)

Association		Variables		
		P–D	*P–V*	*D–V*
Marginal		1.18	2.71	25.99
Partial	Level 1	0.67	2.80	22.04
	Level 2	0.79	3.29	25.90

Note: The value 0.5 was added to each cell frequency before calculation of odds ratios.

The marginal odds ratio for death penalty verdict and defendant's race is 1.18. This means that the estimated odds of the death penalty being imposed were 1.18 times as high for white defendants as for black defendants. The corresponding partial odds ratios mean the following: when the victim was white, the estimated odds of the death penalty being imposed were 0.67 times as high for white defendants as for black defendants. When the victim was black, the estimated odds of the death penalty being imposed were 0.79 times as high for white defendants as for black defendants. Like the sample proportions in Table 3.1, these odds ratios show that the association changes direction when victim's race is controlled.

We can understand the difference between these marginal and partial odds ratios by studying the other odds ratios in Table 3.2. The odds ratios relating victim's race and defendant's race indicate a very strong association between those two variables. To illustrate, the odds of having killed a white are estimated to be 25.99 times higher for white defendants than for black defendants. The odds ratios relating death penalty verdict and victim's race indicate that the death penalty was more likely to be imposed when the victim was white than when the victim was black. The pattern of these two associations suggests that white defendants would be more likely than black defendants to receive the death penalty, which is the result observed in Table 2.1.

Generally, suppose we are interested in the marginal association between two variables X and Y and also in their partial association within fixed levels of a third variable Z. Suppose that Z is associated with both X and Y. Then the partial association between X and Y when Z is controlled may differ from the marginal association, where the effects of Z on the $X-Y$ association have not been controlled.

A condition under which marginal associations and partial associations are identical will be discussed in the next section. For good discussions of the types of bias that can occur when one fails to control relevant variables, see Anderson et al. (1980) and Everitt (1977, pp. 34–36).

3.2. ASSOCIATION AND INTERACTION STRUCTURE

We now give a formal treatment of the ideas introduced in the previous section, and we introduce some structural characteristics that describe various patterns of cell probabilities. Consider a three-dimensional cross-classification of variables (X, Y, Z). Denote the cell probabilities by $\{\pi_{ijk}, i = 1, \ldots, r, j = 1, \ldots, c, k = 1, \ldots, l\}$, where $\sum_i \sum_j \sum_k \pi_{ijk} = 1.0$. We shall first focus on the partial or "conditional" associations between X and Y in partial tables in which the value of Z is fixed.

Conditional Independence

Within a fixed level k of Z, $1 \leq k \leq l$, the conditional $X-Y$ association can be described by the set of odds ratios

$$\frac{\pi_{ijk}\,\pi_{rck}}{\pi_{ick}\,\pi_{rjk}}, \quad 1 \leq i \leq r-1, \qquad 1 \leq j \leq c-1$$

Equivalent information is given by the set of local odds ratios

$$\theta_{ij(k)} = \frac{(\pi_{ijk}\,\pi_{i+1,\,j+1,\,k})}{(\pi_{i,\,j+1,\,k}\,\pi_{i+1,\,j,\,k})}, \qquad 1 \leq i \leq r-1, \qquad 1 \leq j \leq c-1 \quad (3.1)$$

which are especially useful if X and Y are ordinal. Within each level of Z, there are $(r-1)(c-1)$ of these odds ratios, and they determine all odds ratios that can be formed from pairs of rows and pairs of columns. Considered across all l levels of Z, there are $l(r-1)(c-1)$ of these conditional odds ratios. In a similar manner the conditional association between X and Z can be described by $(r-1)(l-1)$ local odds ratios $\{\theta_{i(j)k},\ 1 \leq i \leq r-1$ and $1 \leq k \leq l-1\}$ at each of the c levels of Y, and the conditional association between Y and Z can be described by $(c-1)(l-1)$ local odds ratios $\{\theta_{(i)jk},\ 1 \leq j \leq c-1$ and $1 \leq k \leq l-1\}$ at each of the r levels of X.

The variables X and Y are *conditionally independent at level k of Z* if all $(r-1)(c-1)$ of the $\theta_{ij(k)}$ at that fixed level of Z equal 1.0. They are said to be *conditionally independent given Z* (i.e., at *every* level of Z) if all $l(r-1)(c-1)$ of the $\{\theta_{ij(k)}\}$ equal 1.0.

We showed in Section 3.1 that partial associations can be quite different from marginal associations. If X and Y are conditionally independent, given Z, it does not follow that X and Y are independent in the marginal sense.

The proportions in Table 3.3 illustrate that conditional independence does not imply marginal independence. These proportions show a hypothetical relationship between Y = income (high, low) of assistant professors and X = sex (women, men) at two levels of Z = college in a university (Liberal Arts, Professional). The partial odds ratios relating Y to X at the two levels of Z are

$$\theta_{11(1)} = \frac{(0.18 \times 0.08)}{(0.12 \times 0.12)} = 1.0$$

$$\theta_{11(2)} = \frac{(0.02 \times 0.32)}{(0.08 \times 0.08)} = 1.0$$

Thus income and sex are conditionally independent, given college. The odds ratio for the marginal income-sex table equals $(0.20 \times 0.40)/$

Table 3.3. Hypothetical Distribution Used to Show That Conditional Independence Does Not Imply Marginal Independence

		Income	
College	Sex	Low	High
Liberal Arts	Women	0.18	0.12
	Men	0.12	0.08
Professional	Women	0.02	0.08
	Men	0.08	0.32
Total	Women	0.20	0.20
	Men	0.20	0.40

$(0.20 \times 0.20) = 2.0$, however, so the variables are not independent when college is ignored.

The $X-Y$ marginal odds ratio of 2.0 means that the odds of having a high income are twice as large for men as for women. This association disappears when $Z =$ college is controlled because of the nature of the associations between Z and X and between Z and Y. The $X-Z$ partial odds ratios are $\theta_{1(1)1} = (0.18 \times 0.08)/(0.12 \times 0.02) = 6.0$ and $\theta_{1(2)1} = (0.12 \times 0.32)/(0.08 \times 0.08) = 6.0$. Controlling for income, in other words, the odds of being in the professional college are six times higher for men than for women. The $Y-Z$ partial odds ratios are $\theta_{(1)11} = \theta_{(2)11} = 6.0$. Controlling for sex, the odds of having a high income are six times higher for those in the professional college than for those in the liberal arts college.

In summary, incomes tend to be higher in the professional college (for each sex), and the ratio of men to women is higher in the professional college (at each income level). Hence we expect men to fare better on income (relative to women) when college is ignored rather than when it is controlled. The tendency in the $X-Y$ marginal table for relatively more men than women to have high salaries is "explained" by two factors. First, the professional college employs relatively more men than does the liberal arts college. Second, the professional college has relatively more people at high incomes than does the liberal arts college. For some "real" data for which a marginal association disappears when a third variable is controlled, see Complements 4.2 and 4.3.

Identical Marginal and Partial Associations

Tables 3.1 and 3.3 have shown that we do not usually get the whole picture by analyzing only the marginal tables of multidimensional cross-classifications. Whenever possible, variables should be controlled that may affect the association between the variables of interest. The marginal $X-Y$ association reflects whatever joint effects any other variable has on X and on Y.

We now state an important result that gives conditions under which the $X-Y$ association, as described by the odds ratio, is the same in partial tables as in the marginal table. If these associations are the same, the $X-Y$ association can be studied in a simplified manner by collapsing over the Z dimension. The collapsibility conditions are as follows, where Z may be a single variable or multidimensional:

> Marginal odds ratio measures between X and Y are the same as the corresponding partial odds ratio measures at each level of Z if either or both of the following hold:
>
> **1.** Z and X are conditionally independent, given Y.
> **2.** Z and Y are conditionally independent, given X.

Let $\theta_{ij(k)}$ be as in (3.1), and let θ_{ij}^{XY} denote the corresponding marginal odds ratio defined in (2.15). Then the collapsibility conditions imply that

$$\theta_{ij}^{XY} = \theta_{ij(1)} = \theta_{ij(2)} = \cdots = \theta_{ij(l)}, \qquad 1 \leq i \leq r-1, \qquad 1 \leq j \leq c-1$$

if either or both of the following hold:

$$\theta_{i(j)k} = 1, \qquad 1 \leq i \leq r-1, \qquad 1 \leq j \leq c, \qquad 1 \leq k \leq l-1$$

$$\theta_{(i)jk} = 1, \qquad 1 \leq i \leq r, \qquad 1 \leq j \leq c-1, \qquad 1 \leq k \leq l-1$$

It follows that if the partial association at each level of Z is *not* the same as the marginal association, then Z and X are not conditionally independent, given Y, and also Z and Y are not conditionally independent, given X. The collapsibility conditions will be used often in Section 3.3 to help describe various association patterns.

Interaction

When either collapsibility condition holds, each $X-Y$ partial odds ratio equals the corresponding marginal odds ratio. This implies that for any i and j, the l odds ratios $\{\theta_{ij(k)}, k = 1, \ldots, l\}$ are identical. A three-dimensional table having this property is said to exhibit *no three-factor interaction*. In other words, the absence of three-factor interaction means that the association between X and Y is the same at each level of Z, in the sense that for any i and j,

$$\theta_{ij(1)} = \theta_{ij(2)} = \cdots = \theta_{ij(l)} \tag{3.3}$$

When this property holds, it is also necessarily true that the X–Z association is the same at each level of Y, and that the Y–Z association is the same at each level of X (Complement 3.2).

If any pair of variables is conditionally independent, then there is no three-factor interaction. If there is no three-factor interaction, it is possible that all three pairs of variables are conditionally dependent or that certain pairs of the variables are conditionally independent. The structure of the table is relatively simple in all cases, since the same set of partial odds ratios describes the association between two variables at each level of the third variable. If three-factor interaction exists, no pair of variables can be conditionally independent.

A Hierarchy of Structures for Three Variables

There are various patterns that the partial associations can take in a three-dimensional table. The patterns can be described in terms of structural characteristics such as conditional independence and interaction. For example, a table can be characterized by which (if any) pairs of variables are conditionally independent and by whether there is three-factor interaction.

For three-dimensional tables we will consider a hierarchy of five types of structures, ordered in terms of the extent of association and three-factor interaction:

1. All three pairs of variables are conditionally independent. That is:

    X is independent of Y, given Z.

    X is independent of Z, given Y.

    Y is independent of Z, given X.

2. Two of the pairs of variables are conditionally independent. For example:

    X is independent of Z, given Y.

    Y is independent of Z, given X.

    X and Y are conditionally dependent, given Z.

    Similarly, the sole conditionally dependent pair could be X and Z, or it could be Y and Z.

3. One of the pairs of variables is conditionally independent. For example:

    X is independent of Z, given Y.

    X and Y are conditionally dependent, given Z.

    Y and Z are conditionally dependent, given X.

    Similarly, the sole conditionally independent pair could by X and Y, or it could be Y and Z.

4. None of the pairs of variables is conditionally independent, but there is no three-factor interaction.
5. There is three-factor interaction. Hence all pairs of variables are conditionally dependent, but the association between each pair varies according to the level of the third variable.

In the next section each of these structures will be represented as a loglinear model for the expected cell frequencies.

3.3. LOGLINEAR MODELS FOR THREE DIMENSIONS

The general form given in Section 2.6 for a loglinear model in two dimensions extends directly to higher dimensions. Let $\mu_{ijk} = \log m_{ijk}$. One can express $\log m_{ijk}$ as

$$\log m_{ijk} = \mu + \lambda_i^X + \lambda_j^Y + \lambda_k^Z + \lambda_{ij}^{XY} + \lambda_{ik}^{XZ} + \lambda_{jk}^{YZ} + \lambda_{ijk}^{XYZ} \tag{3.4}$$

where

$$\sum_i \lambda_i^X = \sum_j \lambda_j^Y = \sum_k \lambda_k^Z = \sum_i \lambda_{ij}^{XY} = \sum_j \lambda_{ij}^{XY} = \cdots = \sum_k \lambda_{ijk}^{XYZ} = 0$$

In model (3.4) doubly subscripted terms pertain to partial associations, and the triply subscripted term pertains to three-factor interaction. When certain parameters are equated to zero in (3.4), models are obtained corresponding to the structures discussed in the previous section. We shall now consider those models, and we shall use the collapsibility conditions of the previous section to describe corresponding marginal (two-way) associations. The models are listed in Table 3.4. For simplicity in future discussions each model can be represented by a symbol that lists the superscript for the highest-order term(s) for each variable. The symbol indicates in abbreviated fashion the pairs of variables that are associated. Conditionally dependent variables appear together in the symbol, with no comma between them.

Table 3.4. Some Loglinear Models for Three-Dimensional Tables

Loglinear Model	Symbol
$\log m_{ijk} = \mu + \lambda_i^X + \lambda_j^Y + \lambda_k^Z$	(X, Y, Z)
$\log m_{ijk} = \mu + \lambda_i^X + \lambda_j^Y + \lambda_k^Z + \lambda_{ij}^{XY}$	(XY, Z)
$\log m_{ijk} = \mu + \lambda_i^X + \lambda_j^Y + \lambda_k^Z + \lambda_{ij}^{XY} + \lambda_{jk}^{YZ}$	(XY, YZ)
$\log m_{ijk} = \mu + \lambda_i^X + \lambda_j^Y + \lambda_k^Z + \lambda_{ij}^{XY} + \lambda_{jk}^{YZ} + \lambda_{ik}^{XZ}$	(XY, YZ, XZ)
$\log m_{ijk} = \mu + \lambda_i^X + \lambda_j^Y + \lambda_k^Z + \lambda_{ij}^{XY} + \lambda_{jk}^{YZ} + \lambda_{ik}^{XZ} + \lambda_{ijk}^{XYZ}$	(XYZ)

(X, Y, Z)

Three variables are completely independent if for all i, j, and k, $\pi_{ijk} = \pi_{i++}\pi_{+j+}\pi_{++k}$. On a log scale this can be represented by the model

$$\log m_{ijk} = \mu + \lambda_i^X + \lambda_j^Y + \lambda_k^Z$$

whereby each pair of variables is conditionally independent, given the remaining one. Suppose that this model holds and that a particular $\lambda_i^X > 0$, say. Then at each combination of Y and Z, the log expected frequency at level i of X is larger than the overall mean of the log expected frequencies across all levels of X. When $r = 2$, $\lambda_1^X - \lambda_2^X = 2\lambda_1^X$ is the constant value of $\log(m_{1jk}/m_{2jk})$ at each combination of Y and Z.

Recall that if a variable is conditionally independent of either of the two other variables, then the marginal association for the other two is the same as their partial association. For this model therefore, all three pairs of variables are also marginally independent. For example, the marginal association between X and Y is the same as the partial association between X and Y, given Z, because Z is conditionally independent of X, given Y, or also because Z is conditionally independent of Y, given X. This model is so simple that it rarely gives a good fit in practice.

(XY, Z) or (XZ, Y) or (YZ, X)

The symbol (XY, Z) denotes the model

$$\log m_{ijk} = \mu + \lambda_i^X + \lambda_j^Y + \lambda_k^Z + \lambda_{ij}^{XY},$$

for which the $\{\lambda_{ij}^{XY}\}$ parameters pertain to the partial association between X and Y, given Z. The symbol for the model reflects the conditional dependence of X and Y. For this model X and Z are conditionally independent (given Y), and Y and Z are conditionally independent (given X). If all $\lambda_{ij}^{XY} = 0$, the variables X and Y are also conditionally independent, and the model simplifies to the one just considered, (X, Y, Z).

By the collapsibility conditions, here also all the marginal associations are identical to the partial associations. For example, the $X-Y$ marginal association equals the $X-Y$ partial association, given Z, because Z is independent of X, given Y, or because Z is independent of Y, given X. The $X-Z$ marginal association is the same as the $X-Z$ partial association, given Y (i.e., X and Z are marginally independent), because Y is independent of Z, given X.

There are three separate models at this level of the hierarchy of structures, corresponding to the three possible pairs of variables that can be conditionally dependent. The symbol (XZ, Y) denotes the model such that

X and Z alone are conditionally dependent, and (YZ, X) denotes the model such that Y and Z alone are conditionally dependent.

(XY, YZ) or (XY, XZ) or (XZ, YZ)

There are three separate models in which only one pair of variables is conditionally independent. For example, (XY, YZ) denotes the model

$$\log m_{ijk} = \mu + \lambda_i^X + \lambda_j^Y + \lambda_k^Z + \lambda_{ij}^{XY} + \lambda_{jk}^{YZ}$$

where $\{\lambda_{ij}^{XY}\}$ and $\{\lambda_{jk}^{YZ}\}$ pertain to the X–Y and Y–Z partial associations, respectively. For this model, X and Z (the only variables that do not appear together in the model symbol) are conditionally independent, given Y.

For model (XY, YZ), the marginal X–Y table and the marginal Y–Z table display the same associations as do their corresponding partial tables, because X and Z are conditionally independent. However, the X–Z marginal association may differ from the X–Z partial association (given Y), because Y is conditionally dependent with both X and Z. Thus X and Z may be marginally dependent, even though they are conditionally independent at each level of Y. Table 3.1 in the previous section exhibited this behavior. This is an important model in practice. A noticeable association between X and Z may be shown to be spurious, if the researcher can find a third variable such that the X–Z association disappears when Y is controlled.

(XY, XZ, YZ)

For the model

$$\log m_{ijk} = \mu + \lambda_i^X + \lambda_j^Y + \lambda_k^Z + \lambda_{ij}^{XY} + \lambda_{jk}^{YZ} + \lambda_{ik}^{XZ}$$

partial association terms appear for each pair of variables, so no pair is conditionally independent. This is the special case of the general model in which $\lambda_{ijk}^{XYZ} = 0$ for all i, j, and k, meaning that there is no three-factor interaction. Suppose that $\lambda_{ij}^{XY} > 0$, say, in this model. Then at each level of Z, the expected frequency at combination (i, j) of (X, Y) is larger than would be expected if X and Y were conditionally independent given Z (*i.e.*, m_{ijk} is larger than it would be if all $\lambda_{ij}^{XY} = 0$).

For this model the partial association between any pair of variables is the same at each level of the third variable, but that association may differ from the corresponding marginal association. For example, the partial X–Y association, given Z, may differ from the marginal X–Y association because Z is conditionally dependent with both X and Y.

(XYZ)

The most general model for three variables, model (3.4), allows for three-factor interaction. In other words, each pair of variables may be conditionally dependent, and the association between any pair may depend on the level of the third variable. The generality of this structure is such that it describes the set $S[(XYZ)]$ of all $\{\pi_{ijk}\}$. The set $S[(XY, XZ, YZ)]$ of $\{\pi_{ijk}\}$ that satisfy (XY, XZ, YZ) is a subset of $S[(XYZ)]$, namely the set of $\{\pi_{ijk}\}$ for which all three-factor interaction parameters $\lambda_{ijk}^{XYZ} = 0$. The set $S[(XY, YZ)]$ that satisfies (XY, YZ) is an even smaller subset of $S[(XYZ)]$ that is also a subset of $S[(XY, XZ, YZ)]$. This set $S[(XY, YZ)]$ displays no three-factor interaction *and* conditional independence between X and Z, given Y; that is, all $\lambda_{ijk}^{XYZ} = 0$ and all $\lambda_{ik}^{XZ} = 0$. As Figure 3.1 and Table 3.4 illustrate, there is a nested sequence of sets of $\{\pi_{ijk}\}$ satisfying models at various levels of the hierarchy. For instance,

$$S[(X, Y, Z)] \subset S[(XY, Z)] \subset S[(XY, YZ)] \subset S[(XY, XZ, YZ)] \subset S[(XYZ)]$$

Since $\sum_i \lambda_i^X = 0$ in model (3.4), there are $(r - 1)$ linearly independent $\{\lambda_i^X\}$ parameters. Since $\sum_i \lambda_{ij}^{XY} = \sum_j \lambda_{ij}^{XY} = 0$, there are $(r - 1)(c - 1)$ linearly independent $\{\lambda_{ij}^{XY}\}$ parameters. Similar formulas apply to the other parameters in the models. The total number of linearly independent parameters (including μ) for the general loglinear model (3.4) is

$$1 + (r - 1) + (c - 1) + (l - 1) + (r - 1)(c - 1) + (r - 1)(l - 1) + (c - 1)(l - 1) + (r - 1)(c - 1)(l - 1) = rcl$$

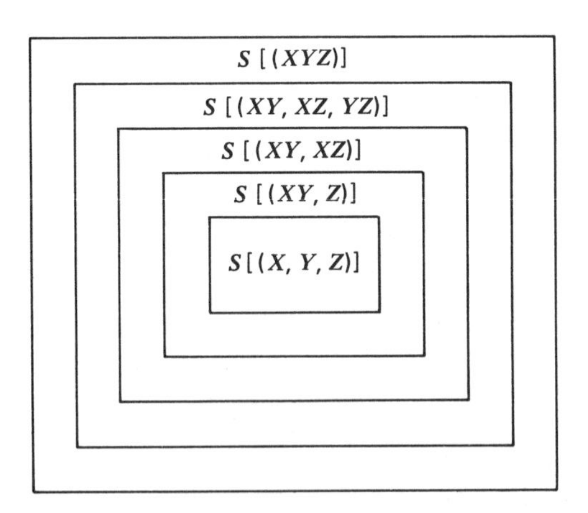

Figure 3.1. Sets of cell proportions satisfying various structures.

which is the total number of cells in the table. Hence the model is saturated.

The loglinear models in Table 3.4 are referred to as *hierarchical* models. This means that whenever the model allows higher-order effects, it also allows lower-order effects composed from the variables. For instance, if λ_{ijk}^{XYZ} is in the model, then so must be λ_{ij}^{XY}, λ_{ik}^{XZ}, and λ_{jk}^{YZ}. Similarly, if λ_{ij}^{XY} is in the model, then λ_i^X and λ_j^Y must also be included. An example of a non-hierarchical model in two dimensions is $\log m_{ij} = \mu + \lambda_j^Y + \lambda_{ij}^{XY}$. This model permits association between X and Y but forces the average log expected frequency across columns to be the same in every row. Since nonhierarchical models are not very meaningful in most applications, they are not considered in this text.

3.4. LOGLINEAR MODELS FOR HIGHER DIMENSIONS

We have seen that the structure of a three-dimensional cross-classification table is more complicated than that of a two-way table because of the partial association and three-factor interaction patterns that must be considered. Once we understand the various models for three-dimensional tables, however, it is relatively easy to understand those for higher-dimensional tables. The main difficulty encountered as the dimensionality increases is the tremendous increase in the number of different models needed to describe all the possible association patterns and all the possible higher-order interactions. For the four-dimensional table, for example, there are 112 different hierarchical loglinear models that are more complex than the mutual independence model. The models are easy to fit using available computer software, but they are not all easy to interpret. Also, as the number of dimensions increases, the number of cells increases dramatically. Unless there is a very large sample size there will be many zero cell counts, and this may affect the asymptotic chi-squared distribution (see Bishop et al. 1975, pp. 59, 115–119).

To illustrate models for higher dimensions, we consider some possible models for a four-dimensional cross-classification of variables W, X, Y, and Z. Usually the simplest model of interest is that whereby the variables are mutually independent, denoted by (W, X, Y, Z). This model rarely provides an adequate fit, however. The models that yield the simplest interpretations are those having no three-factor interaction. Such models are nested within the model

$$\log m_{hijk} = \mu + \lambda_h^W + \lambda_i^X + \lambda_j^Y + \lambda_k^Z + \lambda_{hi}^{WX} + \lambda_{hj}^{WY} + \lambda_{hk}^{WZ} + \lambda_{ij}^{XY} + \lambda_{ik}^{XZ} + \lambda_{jk}^{YZ}$$

denoted by (WX, WY, WZ, XY, XZ, YZ). For this model each of the six pairs of variables is conditionally dependent given the remaining two variables. Conditional independences correspond to the absence of certain two-factor association terms. If $\lambda_{ij}^{XY} = 0$ for all i and j, for instance, then X and Y are conditionally independent within each combination of levels of W and Z.

There are also several models that exhibit some type of three-factor interaction. In the model denoted by (WXY, WZ, XZ, YZ), for instance, each pair of variables is conditionally dependent, but (within each level of Z) the association between W and X or between W and Y or between X and Y varies across the levels of the remaining one of the three variables. The most complex unsaturated model (WXY, WXZ, WYZ, XYZ) has all possible three-factor interactions.

The most general model (the saturated one) for a four-dimensional table is the one ($WXYZ$) corresponding to four-factor interaction. This means, for instance, that the three-factor interaction among W, X, and Y varies among the levels of Z; that is, the way in which the association between X and Y varies across the levels of W is itself different within the various categories of Z.

In fitting higher-dimensional models to data, it is clearly advantageous for interpretive purposes to try to find a relatively simple one that fits well—for example, one that has no or very few three-factor or higher-order interaction terms.

NOTES

Section 3.1

1. A variable Z such that the marginal association between X and Y differs from the partial association controlling for Z is sometimes referred to as a *confounding* variable.

Section 3.2

2. The term "second-order interaction" is sometimes used synonymously for "three-factor interaction." Similarly, association between two variables is often referred to as "two-factor interaction" or "first-order interaction."

3. Fienberg (1980, p. 50) gives a condition under which the $X-Y$ marginal association is the same as the $X-Y$ partial association, even though all three pairs of variables are conditionally dependent. See also Whittemore (1978).

Section 3.3

4. Loglinear models can also be given a geometric representation. For a two-dimensional $r \times c$ table, for instance, the set of all possible $\{\log m_{ij}\}$ is rc-dimensional space. This space therefore corresponds to the general model for all its possible parameter values. Unsaturated models constrain the $\{\log m_{ij}\}$ to some p-dimensional linear manifold in that Euclidean space, with $p < rc$. For example, for the independence model $\log m_{ij} = \mu + \lambda_i^X + \lambda_j^Y$, the $\{\log m_{ij}\}$ are constrained to fall on a $p = (r + c - 1)$-dimensional linear manifold. More complex models provide greater flexibility for fitting data, since the linear manifold on which the $\{\log m_{ij}\}$ can fall has a greater number of dimensions. Haberman (1974a) presented a theoretical development of loglinear models based largely on geometrical ideas.
5. In this book we do not discuss the class of models, even simpler than the mutual independence models, in which some marginal effects are omitted. In two dimensions, for instance, the model $\log m_{ij} = \mu + \lambda_i^X$ implies equality of all c expected cell counts in a given row. Models of this sort are referred to as *noncomprehensive* models.

Section 3.4

6. Upton (1978, pp. 74–80) provides a useful illustration of the process of fitting a loglinear model to a five-dimensional table.

COMPLEMENTS

1. Simpson's paradox states that for three events A, B, and C, it is possible that

$$P(\mathrm{A} \mid \mathrm{B}) > P(\mathrm{A} \mid \mathrm{B}^c)$$

even if

$$P(\mathrm{A} \mid \mathrm{BC}) < P(A \mid B^c\mathrm{C}) \quad \text{and} \quad P(\mathrm{A} \mid \mathrm{BC}^c) < P(\mathrm{A} \mid \mathrm{B}^c\mathrm{C}^c).$$

See Yule (1903), Simpson (1951), Blyth (1972). Show how this paradox, applied to $2 \times 2 \times 2$ cross-classification tables, implies that the marginal $X-Y$ association may have a different direction than the partial $X-Y$ association, controlling for Z. Note that Table 3.1 satisfies Simpson's paradox in terms of sample proportions if A = {death penalty verdict "yes"}, B = {"white" defendant}, and C = {"white" victim}.

2. Show that if

$$\theta_{ij(1)} = \cdots = \theta_{ij(l)}, \quad 1 \le j \le r - 1, 1 \le j \le c - 1$$

then necessarily

$$\theta_{i(1)k} = \cdots = \theta_{i(c)k}, \quad 1 \le i \le r - 1, 1 \le k \le l - 1$$

and

$$\theta_{(1)jk} = \cdots = \theta_{(r)jk}, \quad 1 \le j \le c - 1, 1 \le k \le l - 1$$

Note that this property gives an argument for the use of the odds ratio as a measure of association, in the sense that other measures of association (that are not functions of the odds ratio) do not share this property.

3. Give an example of cell proportions for a $2 \times 2 \times 2$ table that satisfy model

a. (X, Y, Z)

b. (XY, Z)

c. (XY, YZ)

d. (XY, XZ, YZ)

e. (XYZ)

4. Note that there are four terms in the general model (2.23) for two-dimensional tables and eight terms in the general model (3.4) for three-dimensional tables. Show that the general model in d dimensions has 2^d terms.

5. Define partial association measures analogous to (3.1) but expressed in terms of local-global odds ratios. Does the property described in Complement 3.2 hold for this type of odds ratio?

CHAPTER 4

Data Analysis Using Loglinear Models

Chapter 3 introduced loglinear models that correspond to various association and interaction patterns in a multidimensional table. The discussion emphasized concepts, rather than data analysis. This chapter illustrates the use of loglinear models for analyzing categorical data. First, for several models we show how to calculate estimates of expected frequencies that satisfy the model. Chi-squared statistics compare these estimates to the observed data to test whether the model fits adequately. Section 4.2 deals with comparisons of models through odds ratio descriptions, estimates of model parameters, and formal significance tests. Section 4.3 discusses general guidelines for selecting a model to describe a set of data. Section 4.4 gives an iterative method of fitting those models for which estimates of expected frequencies do not have closed-form expressions. The final section discusses reasons that these standard loglinear models may not be appropriate for analyzing ordinal data.

4.1. FITTING MODELS FOR THREE VARIABLES

If we assume that a set of variables satisfies a given loglinear model, then we can use sample data to get estimates of cell proportions or expected frequencies for that model. The estimated expected frequencies can be compared to the observed frequencies using chi-squared statistics in order to evaluate the goodness of fit of the model. This was shown in Section 2.2 for the independence model in a two-way table. In this section, we will illustrate this process with models for three-dimensional tables.

Expected Frequency Estimates

Suppose that the sample arises from full multinomial sampling with true cell probabilities $\{\pi_{ijk}\}$ for the cross-classification of X, Y, and Z, and let $\{m_{ijk}\}$ be the expected cell frequencies; that is, m_{ijk} is the expected value of the marginal binomial random variable for the number of observations in cell (i, j, k). Since that binomial distribution is based on n independent trials and "success probability" π_{ijk}, we have $m_{ijk} = n\pi_{ijk}$.

For each model the maximum likelihood (ML) estimates of the $\{m_{ijk}\}$ depend on the cell counts only through certain "sufficient" statistics. The minimal sufficient statistics are the marginal distributions corresponding to the terms in the symbol for the model (see Note 4.1). For the model (XY, XZ, YZ) having all two-factor dependencies, for instance, the minimal sufficient statistics are the two-dimensional marginal frequencies $\{n_{ij+}\}$, $\{n_{i+k}\}$, and $\{n_{+jk}\}$. Table 4.1 gives the minimal sufficient statistics for models at the various levels of the hierarchy presented in Section 3.3. Notice that more of the sample information is used as the model gets more complicated.

The ML estimates $\{\hat{m}_{ijk}\}$ of the expected cell frequencies can be used to test the null hypothesis that the population cell proportions satisfy the assumed model. We shall use the likelihood-ratio statistic

$$G^2 = 2 \sum \sum \sum n_{ijk} \log \left(\frac{n_{ijk}}{\hat{m}_{ijk}} \right)$$

though the Pearson statistic can also be used.

For each model Table 4.2 contains expressions for the $\{\hat{m}_{ijk}\}$ and for the degrees of freedom for the chi-squared goodness-of-fit test. That table also expresses the cell probabilities $\{\pi_{ijk}\}$ for each model in terms of the marginal probabilities whose sample values are sufficient statistics. These expressions help to motivate the formulas for the $\{\hat{m}_{ijk}\}$. For instance, if the complete independence model (X, Y, Z) holds, then

$$\pi_{ijk} = \pi_{i++}\pi_{+j+}\pi_{++k}, \qquad 1 \le i \le r, \qquad 1 \le j \le c, \qquad 1 \le k \le l$$

Table 4.1. Minimal Sufficient Statistics for Estimating Expected Frequencies

Model	Minimal Sufficient Statistics
(X, Y, Z)	$\{n_{i++}\}$, $\{n_{+j+}\}$, $\{n_{++k}\}$
(XY, Z)	$\{n_{ij+}\}$, $\{n_{++k}\}$
(XY, YZ)	$\{n_{ij+}\}$, $\{n_{+jk}\}$
(XY, XZ, YZ)	$\{n_{ij+}\}$, $\{n_{i+k}\}$, $\{n_{+jk}\}$

Table 4.2. Expected Frequency Estimates and Degrees of Freedom for Goodness-of-Fit Test

Model	Probabilistic Form	Expected Frequency	Degrees of Freedom	Number of Models of This Type
(X, Y, Z)	$\pi_{ijk} = \pi_{i++}\pi_{+j+}\pi_{++k}$	$\hat{m}_{ijk} = n_{i++}n_{+j+}n_{++k}/n^2$	$rcl - r - c - l + 2$	1
(XY, Z)	$\pi_{ijk} = \pi_{ij+}\pi_{++k}$	$\hat{m}_{ijk} = n_{ij+}n_{++k}/n$	$(l - 1)(rc - 1)$	3
(XY, YZ)	$\pi_{ijk} = \left(\frac{\pi_{ij+}}{\pi_{+j+}}\right)\left(\frac{\pi_{+jk}}{\pi_{+j+}}\right)$	$\hat{m}_{ijk} = n_{ij+}n_{+jk}/n_{+j+}$	$c(r - 1)(l - 1)$	3
(XY, XZ, YZ)	$\theta_{ij(1)} = \cdots = \theta_{ij(l)}$	See Section 4.4	$(r - 1)(c - 1)(l - 1)$	1
(XYZ)		$\hat{m}_{ijk} = n_{ijk}$	0	1

Note: Formulas for the other models at the second and third levels of the hierarchy are obtained by symmetry. For example, for the model (XZ, Y), $\hat{m}_{ijk} = n_{i+k}n_{+j+}/n$ and df $= (c - 1)(rl - 1)$.

The ML estimates of marginal probabilities such as $\{\pi_{i++}\}$ are the sample proportions $\{p_{i++} = n_{i++}/n\}$, so the ML estimate of π_{ijk} is

$$\hat{\pi}_{ijk} = p_{i++}\, p_{+j+}\, p_{++k} = \frac{n_{i++}\, n_{+j+}\, n_{++k}}{n^3}$$

The corresponding estimate of the expected frequency m_{ijk} is

$$\hat{m}_{ijk} = n\hat{\pi}_{ijk} = \frac{n_{i++}\, n_{+j+}\, n_{++k}}{n^2}$$

No closed-form expression can be given for $\{\hat{m}_{ijk}\}$ for model (XY, XZ, YZ), and we shall show how the estimates can be obtained for that case in Section 4.4.

The number of degrees of freedom for the chi-squared test for a model is the difference in dimension between the alternative and null hypotheses. This equals the difference between the number of parameters that determine the $\{\pi_{ijk}\}$ in the general case and in the case that the model holds. In the general case (i.e., under the alternative hypothesis) the only constraint is $\sum_i \sum_j \sum_k \pi_{ijk} = 1$, so there are $rcl - 1$ linearly independent parameters. Given $rcl - 1$ of the cell proportions, the remaining one can be determined, since they sum to 1.0. In the case of the independence model (X, Y, Z), the $\{\pi_{ijk} = \pi_{i++}\, \pi_{+j+}\, \pi_{++k}\}$ are determined by $r - 1$ of the $\{\pi_{i++}\}$ (since $\sum_i \pi_{i++} = 1$), $c - 1$ of the $\{\pi_{+j+}\}$, and $l - 1$ of the $\{\pi_{++k}\}$, a total of $r + c + l - 3$ parameters. The degrees of freedom for the chi-squared test equal

$$(rcl - 1) - (r + c + l - 3) = rcl - r - c - l + 2$$

as reported in Table 4.2. The degrees of freedom also equal the number of cells in the table minus the number of linearly independent parameters in the model. In other words, the degrees of freedom correspond to the number of linearly independent parameters in the most general model that are set equal to zero to obtain that model. For the complete independence model

$$\log m_{ijk} = \mu + \lambda_i^X + \lambda_j^Y + \lambda_k^Z$$

for instance, $\text{df} = rcl - [1 + (r - 1) + (c - 1) + (l - 1)] = rcl - r - c - l + 2$.

A model is saturated when it has as many parameters as there are cells in the table, so that $\text{df} = 0$. The saturated model is the most general model, and it gives a perfect fit to the observed data. For model (XYZ), for instance, the estimated expected frequencies are simply the observed cell counts.

The estimated expected frequencies are identical to the observed data in the marginal distributions corresponding to the minimal sufficient statistics.

For the model (XY, Z), for instance, from Table 4.2,

$$\hat{m}_{ij+} = \sum_k \hat{m}_{ijk} = \sum_k \frac{n_{ij+} n_{++k}}{n} = \left(\frac{n_{ij+}}{n}\right) \sum_k n_{++k} = n_{ij+}$$

and

$$\hat{m}_{++k} = \sum_i \sum_j \frac{n_{ij+} n_{++k}}{n} = \left(\frac{n_{++k}}{n}\right) \sum_i \sum_j n_{ij+} = n_{++k}$$

The estimated expected frequencies can be regarded as smoothed versions of the sample frequencies that match them in certain marginal distributions but satisfy certain independence or no three-factor interaction patterns.

The $\{\hat{m}_{ijk}\}$ are the same for the Poisson sampling model discussed in Section 2.2 as they are for the full multinomial sampling model (see Bishop et al. 1975, pp. 446–448). For independent multinomial sampling we should only fit those models that include terms for the marginal frequencies that are fixed by the sampling design. For example, suppose we select an independent multinomial sample on Y within each combination of levels of X and Z. Then the sample sizes $\{n_{i+k}\}$ are not random but rather are fixed by the sampling design, so the $\{\hat{m}_{ijk}\}$ should satisfy $\hat{m}_{i+k} = n_{i+k}$ for all i and k. Hence the term λ_{ik}^{XZ} should appear in the loglinear model. Subject to this restriction, the $\{\hat{m}_{ijk}\}$ for independent multinomial sampling are the same as for full multinomial sampling. See the example in Section 4.5 for further discussion of this point.

Death Penalty Example

In Chapter 3 (Table 3.1) we introduced Radelet's $2 \times 2 \times 2$ cross-classification for the effect of defendant's race and victim's race on the imposition of the death penalty. These data will now be used to illustrate loglinear models for three variables. In the symbols for the models, D represents defendant's race, V represents victim's race, and P represents death penalty verdict.

After calculating the $\{\hat{m}_{ijk}\}$ and df values using the formulas in Table 4.2, we can obtain the G^2 and P-values for testing goodness of fit. These are given in Table 4.3 for the hierarchical models we have discussed. There are widely available computer programs that calculate the $\{\hat{m}_{ijk}\}$ and associated G^2 values for loglinear models (see Appendix D). We shall take a closer look at the $\{\hat{m}_{ijk}\}$ for certain models shortly.

The G^2 statistics and associated P-values in Table 4.3 indicate that the (D, V, P), (VP, D), (DP, V), and (VP, DP) models represent the observed data very poorly. The common feature of these models is that they omit the

Table 4.3. Goodness-of-Fit Tests for Loglinear Models Relating Death Penalty Verdict (P), Defendant's Race (D), and Victim's Race (V)

Model	G^2	df	P-value
(D, V, P)	137.93	4	0.000
(VP, D)	131.68	3	0.000
(DP, V)	137.71	3	0.000
(DV, P)	8.13	3	0.043
(DP, VP)	131.46	2	0.000
(DP, DV)	7.91	2	0.019
(VP, DV)	1.88	2	0.390
(DP, VP, DV)	0.70	1	0.402
(DVP)	0	0	—

Source: Data from Radelet (1981). See Table 3.1.

$D-V$ association. Hence there is evidence of an important association between defendant's race and victim's race. Of the remaining four unsaturated models, (VP, DV) and (VP, DP, DV) are the only ones that give an adequate fit according to a formal 0.05-level significance test.

In the model (VP, DP, DV) all pairs of variables are conditionally dependent, but there is no three-factor interaction. For that model the death penalty verdict is associated both with defendant's race and with victim's race. According to the simpler model (VP, DV), the death penalty verdict is independent of defendant's race, given victim's race. Remember that this model does not imply that the marginal $D-P$ table will display independence. For this model victim's race is conditionally dependent with death penalty verdict and with defendant's race, so the marginal $D-P$ association may differ from the partial $D-P$ association given V.

Table 4.4 lists the estimated expected frequencies for the best-fitting model at each level of the hierarchy. Let $\hat{m}_{ijk}$ denote the estimated expected frequency at level i of defendant's race, level j of victim's race, and level k of death penalty verdict. Then, for example, for model (D, V, P),

$$\hat{m}_{111} = \frac{n_{1++}n_{+1+}n_{++1}}{n^2} = \frac{(160 \times 214 \times 36)}{(326)^2} = 11.6$$

for model (P, DV),

$$\hat{m}_{111} = \frac{n_{11+}n_{++1}}{n} = \frac{(151 \times 36)}{326} = 16.7$$

Table 4.4. Estimated Expected Frequencies for Five Loglinear Models Fitted to the Death Penalty Data

Defendant's Race	Victim's Race	Death Penalty	Model				
			(D, V, P)	(P, DV)	(VP, DV)	(DP, VP, DV)	(DVP)
White	White	Yes	11.60	16.68	21.17	18.67	19
		No	93.43	134.32	129.83	132.32	132
	Black	Yes	6.07	0.99	0.48	0.33	0
		No	48.90	8.01	8.52	8.68	9
Black	White	Yes	12.03	6.96	8.83	11.32	11
		No	96.94	56.04	54.17	51.70	52
	Black	Yes	6.30	11.37	5.52	5.68	6
		No	50.73	91.63	97.48	97.30	97
	G^2		137.9	8.1	1.9	0.7	0
	df		4	3	2	1	0
	P-value		0.00	0.04	0.39	0.40	—

for model (VP, DV),

$$\hat{m}_{111} = \frac{n_{11+} n_{+11}}{n_{+1+}} = \frac{(151 \times 30)}{214} = 21.2$$

and for model (DVP), the saturated one,

$$\hat{m}_{111} = n_{111} = 19.0$$

The estimated expected frequencies for models (VP, DV) and (VP, DP, DV) are uniformly quite close to the observed frequencies.

Notice that $\hat{m}_{121} > 0$ for each unsaturated model even though $n_{121} = 0$. If there are "too many" zero cell frequencies in a table, it may be impossible to fit certain models. If the zeroes occur in a pattern such that necessarily some $\hat{m}_{ijk} = n_{ijk}$ exactly, one must adjust the degrees of freedom for the asymptotic chi-squared distribution (see Bishop et al. 1975, pp. 59, 115–119). A zero that occurs in a cell where it is theoretically impossible to have any observations is called a *structural zero*. A zero that occurs in a cell for which $m_{ijk} > 0$, so that $n_{ijk} = 0$ is due to the smallness of the sample size, is called a *sampling zero*. To eliminate difficulties caused by sampling zeroes, one can add a small constant to each cell in the table. Some authors recommend routinely adding 0.5 (see Goodman 1971), whereas others suggest that the constant should depend on the data and possibly on a prior distribution for the cell probabilities (see Note 4.3).

4.2. COMPARISON OF MODELS

In the previous section, formal goodness-of-fit tests indicated that the models (VP, DV) and (VP, DP, DV) give adequate fits to the death penalty data. We will now explore further the nature of the associations represented by these models and show how to choose between two models when one is a special case of the other.

Behavior of Odds Ratios

In Chapter 3 we discussed the behavior of marginal and partial associations for various three-dimensional models. As an illustration of the various patterns of association, Table 4.5 presents estimated odds ratios for all marginal and partial associations for the models estimated in Table 4.4. For example, the entry 1.0 for the partial association for model (VP, DV) is the common value of the D–P partial odds ratios of $\{\hat{m}_{ijk}\}$ at the two levels of

Table 4.5. Summary of Estimated Odds Ratios

Model	Partial Associations			Marginal Associations		
	D–P	V–P	D–V	D–P	V–P	D–V
(D, V, P)	1.0	1.0	1.0	1.0	1.0	1.0
(P, DV)	1.0	1.0	27.4	1.0	1.0	27.4
(VP, DV)	1.0	2.9	27.4	1.65	2.9	27.4
(DP, VP, DV)	0.6	3.7	28.7	1.2	2.9	27.4
(DVP) Level 1	0.68	∞	∞	1.2	2.9	27.4
	(0.67)	(2.80)	(22.04)	(1.18)	(2.71)	(25.99)
Level 2	0	3.42	27.36			
	(0.79)	(3.29)	(25.90)			

Note: Values in parentheses for model (DVP) are obtained after adding 0.5 to each cell.

V; that is,

$$1.0 = \frac{(21.17 \times 54.17)}{(129.83 \times 8.83)} = \frac{(0.48 \times 97.48)}{(8.52 \times 5.52)}$$

The entry of 1.65 for the marginal D–P association for that same model is the corresponding odds ratio of estimated expected frequencies for the marginal D–P table; that is,

$$1.65 = \frac{(21.17 + 0.48)(54.17 + 97.48)}{(129.83 + 8.52)(8.83 + 5.52)}$$

The odds ratios for the observed data are those reported for model (DVP), since there $\hat{m}_{ijk} = n_{ijk}$. For that model the entry $\hat{m}_{121} = 0$ causes some of the partial odds ratios to equal 0 or ∞. Hence for model (DVP) we have also reported the odds ratios obtained after adding 0.5 to each cell count.

For each model the estimated odds ratios reflect the relationship between marginal and partial associations described by the collapsibility conditions in Section 3.2.

(D, V, P)

All partial odds ratios equal one and are necessarily the same as the marginal odds ratios, since the collapsibility conditions are fulfilled.

(P, DV)

Only the D–V odds ratios do not equal one. All partial odds ratios are necessarily the same as the marginal odds ratios. The estimated marginal

$D-V$ odds ratio is the same as for the observed data, since $\{n_{ij+}\}$ is part of the minimal sufficient set of statistics for this structure and thus all $\hat{m}_{ij+} = n_{ij+}$.

(VP, DV)

Only the $D-P$ partial odds ratios equal one. The $V-P$ and $D-V$ partial odds ratios are necessarily the same as the marginal odds ratios. This is not true of the $D-P$ partial and marginal odds ratios, since V is not conditionally independent of P or D. The estimated marginal $V-P$ and $D-V$ odds ratios are the same as for the observed data, since all $\hat{m}_{ij+} = n_{ij+}$ and $\hat{m}_{+jk} = n_{+jk}$.

(DP, VP, DV)

All pairs of variables are conditionally dependent. Thus none of the partial odds ratios equal one, and none of the partial odds ratios need be the same as the corresponding marginal odds ratios. The estimated marginal odds ratios equal those for the observed data, since all $\hat{m}_{ij+} = n_{ij+}$, $\hat{m}_{i+k} = n_{i+k}$, and $\hat{m}_{+jk} = n_{+jk}$.

(DVP)

Since this model allows three-factor interaction, the two partial odds ratios for a given pair of variables are no longer equal. They *are* close (after 0.5 is added to each cell), however, which indicates why the model (*DP*, *VP*, *DV*) fits so well.

The values of these odds ratios are useful for interpreting the associations. For instance, consider the ones for model (*DP*, *VP*, *DV*), which fit the observed cell counts well but provide a smoothing of them. The $D-P$ odds ratios means that the estimated odds of the death penalty verdict "yes" are:

1. 1.2 times as high for a white defendant as for a black defendant.
2. 0.6 times as high for a white defendant as for a black defendant, within each level of victim's race.

The reversal in the nature of the $D-P$ association when V is controlled was noted and explained for the observed data in Section 3.1.

Conditional Tests for Nested Models

An important property of the likelihood-ratio statistic is its monotone decreasing behavior as terms are added to a model. To illustrate, notice in Table 4.4 that

$$0 = G^2[(DVP)] \leq G^2[(DP, VP, DV)] \leq G^2[(VP, DV)] \leq G^2[(P, DV)]$$
$$\leq G^2[(D, V, P)]$$

As mentioned in Section 2.3, the difference in G^2 values can be used to compare two nested models. Suppose that model (2) is a special case of model (1). Given that the variables satisfy model (1), we can test whether they also satisfy the simpler model (2) by testing the null hypothesis that the extra terms in model (1) [i.e., those lambda parameters not in model (2)] equal zero. The appropriate test statistic is

$$G^2[(2) \mid (1)] = G^2[(2)] - G^2[(1)]$$

Under the null hypothesis this test statistic has an asymptotic chi-squared distribution with degrees of freedom equal to the difference between the degrees of freedom for model (2) and for model (1). This test is useful for comparing two models that both seem to give adequate fits to a data set and also for testing a hypothesis under the assumption that a certain structure (e.g., no three-factor interaction) holds.

We have seen that model (DP, VP, DV) gives a good fit to the death penalty data. However, we have also seen that the simpler model (VP, DV) may be adequate for describing the data. To test formally whether (VP, DV) fits, given that the variables satisfy model (DP, VP, DV), we test the hypothesis H_0: $\lambda_{ik}^{DP} = 0$ for all i and k for the model

$$\log m_{ijk} = \mu + \lambda_i^D + \lambda_j^V + \lambda_k^P + \lambda_{ij}^{DV} + \lambda_{jk}^{VP} + \lambda_{ik}^{DP}$$

The test statistic is

$$G^2[(VP, DV) \mid (DP, VP, DV)] = G^2[(VP, DV)] - G^2[(DP, VP, DV)]$$
$$= 1.88 - 0.70 = 1.18,$$

based on df $= 2 - 1 = 1$. This gives a test of the null hypothesis that death penalty verdict and defendant's race are conditionally independent (given victim's race) under the assumption of no three-factor interaction. The null hypothesis would not be rejected in this case; that is, the model (VP, DV) does not give a significantly poorer fit than the model (DP, VP, DV).

The *unconditional* goodness-of-fit test of model (VP, DV), based on $G^2[(VP, DV)]$, does not assume an absence of three-factor interaction. In it, the null hypothesis corresponds to the statement that the λ_{ik}^{DP} and λ_{ijk}^{DVP}

terms in the saturated model equal zero. The test statistic in that case can be expressed trivially in the conditional form $G^2[(VP, DV)] = G^2[(VP, DV) \mid (DVP)] = G^2[(VP, DV)] - G^2[(DVP)]$.

Parameter Estimates

Estimates of the lambda parameters in loglinear models can be obtained by substituting the estimated expected frequencies into defining formulas for the lambdas. The lambda estimates can be requested in most computer packages that fit loglinear models. For a multidimensional table $\{n_u\}$ having arbitrary number of dimensions, denote an arbitrary model parameter estimate by $\hat{\lambda}$. It can be shown that:

1. For the general loglinear model $\hat{\lambda}$ is a linear combination of the $\{\log n_u\}$, say $\sum a_u \log n_u$ for some set of constants $\{a_u\}$.
2. The estimated asymptotic variance of $\hat{\lambda}$ has the form $\hat{\sigma}^2(\hat{\lambda}) = \sum a_u^2/\hat{m}_u$.
3. $\hat{\lambda}$ is asymptotically normally distributed.

Properties 2 and 3 imply that for large samples $z = (\hat{\lambda} - \lambda)/\hat{\sigma}(\hat{\lambda})$ can be treated like a standard normal deviate in making inferences about λ.

The magnitudes of the lambda estimates are easiest to interpret when the relevant variables have only two levels. For example, for the saturated model for the $2 \times 2 \times 2$ table,

$$\begin{aligned}
\hat{\lambda}_{111}^{XYZ} &= \left(\frac{1}{8}\right) \log \left[\frac{(n_{111}n_{221}n_{122}n_{212})}{(n_{121}n_{211}n_{112}n_{222})}\right] \\
&= \left(\frac{1}{8}\right) \log \left(\frac{\hat{\theta}_{11(1)}}{\hat{\theta}_{11(2)}}\right) \\
&= \left(\frac{1}{8}\right) \log \left(\frac{\hat{\theta}_{1(1)1}}{\hat{\theta}_{1(2)1}}\right) \\
&= \left(\frac{1}{8}\right) \log \left(\frac{\hat{\theta}_{(1)11}}{\hat{\theta}_{(2)11}}\right)
\end{aligned} \tag{4.1}$$

In that case $\hat{\lambda}_{111}^{XYZ} = 0$ (and hence the other $\hat{\lambda}_{ijk}^{XYZ} = 0$) if and only if the sample conditional odds ratio relating a pair of variables is identical at both levels of the third variable. The $\{\hat{m}_{ijk}\}$ for any unsaturated model in Table 3.4 are such that the corresponding estimate

$$\hat{\lambda}_{111}^{XYZ} = \left(\frac{1}{8}\right) \log \left(\frac{\hat{m}_{111}\hat{m}_{221}\hat{m}_{122}\hat{m}_{212}}{\hat{m}_{121}\hat{m}_{211}\hat{m}_{112}\hat{m}_{222}}\right) = 0$$

For the $2 \times 2 \times I$ table, any model that assumes no three-factor interaction satisfies

$$\hat{\lambda}_{11}^{XY} = \left(\frac{1}{4}\right) \log \left(\frac{\hat{m}_{11k}\hat{m}_{22k}}{\hat{m}_{12k}\hat{m}_{21k}}\right), \qquad k = 1, \ldots, l \tag{4.2}$$

Thus, as in the two-dimensional case [see (2.24)], the association parameter is directly related to the odds ratio measure. If a model fits, the signs of the estimates of the highest-order lambda parameters indicate whether the estimated expected frequencies in certain cells are higher or lower than they would be expected to be if those effects were absent.

For Radelet's death penalty data, Table 4.6 contains parameter estimates for model (VP, DV), which was the simplest one to fit the data adequately, and model (DVP). The standardized values given in that table are the parameter estimates divided by their asymptotic standard errors. Under the null hypothesis that $\lambda = 0$, the standardized value has an asymptotic standard normal distribution and hence can be compared to entries in a normal table.

For the saturated model the standardized values of $\hat{\lambda}_{11}^{DP}$ and $\hat{\lambda}_{111}^{DVP}$ are quite small, so it is not surprising that the model (VP, DV) fits well. The positive value of $\hat{\lambda}_{11}^{VP}$ for model (VP, DV) indicates that when the victim was white, the death penalty was imposed relatively more often than would be expected if those variables were independent given defendant's race. From (4.2), $\exp(4\hat{\lambda}_{11}^{VP}) = 2.9$ for this model, and this is the estimated VP odds ratio reported for it in Table 4.5.

Table 4.6. Estimates of Parameters for Models (DVP) and (VP, DV) Fit to Death Penalty Data

	Model (DVP)		Model (VP, DV)	
Parameter	Estimate	Std Value	Estimate	Std Value
μ	2.78		2.72	
λ_1^D	−0.43		−0.39	
λ_1^V	0.78		0.80	
λ_1^P	−1.14		−1.17	
λ_{11}^{DV}	0.79	4.07	0.83	8.75
λ_{11}^{VP}	0.28	1.42	0.26	2.28
λ_{11}^{DP}	−0.08	−0.41	0	—
λ_{111}^{DVP}	−0.02	−0.10	0	—

Note: Other parameter estimates are determined by the parameter constraints. For instance, $\lambda_{12}^{DV} = \lambda_{21}^{DV} = -\lambda_{11}^{DV} = -\lambda_{22}^{DV}$. For the model ($DVP$), $\hat{m}_{121} = 0$, so 0.5 was added to each cell count before the parameter estimates were calculated.

4.3. MODEL-BUILDING

The selection of a loglinear model for describing a data set becomes more complicated as the number of variables increases, since there is a rapid increase in possible associations and interactions. It is usually too cumbersome to fit all possible models when the number of dimensions exceeds three, and it is useful to have guidelines for selecting a model. As in the choice of a regression model, we must balance two goals. We want a complex enough model so that we obtain a reasonably good fit to the data. On the other hand, we want a model that is relatively simple to interpret, one that smooths rather than overfits the data.

Standardized Parameter Estimates

Usually certain terms are included in the loglinear model to help answer the questions posed in the research study. To get some indication of which additional terms are needed in the model, one can fit the saturated model and observe which lambda estimates have large standardized values. A natural starting point for a model is a hierarchical one containing all lambda terms that have at least one standardized estimate $z = \hat{\lambda}/\hat{\sigma}(\hat{\lambda})$ exceeding in absolute value a certain number, say 2.0. If that model fits well, it may be possible to simplify it and still maintain an adequate model. If it does not fit well, additional parameters having moderate standardized estimates can be added until the fit is adequate.

According to this strategy, inspection of the lambda estimates in Table 4.6 for the saturated model for the death penalty data leads directly to the model (VP, DV), if 2.0 is the critical value for $|z|$.

Partitioning Chi-Squared

Often research considerations suggest that a small set of models are of special interest. We may wish to compare two models that differ only by the inclusion of a certain association term, for instance. If the models to be compared form a nested set, then conditional test statistics can be formed that partition the chi-squared statistic for the simplest model and that test a sequence of hypotheses about associations.

To illustrate, consider the set of nested models $\{(XY, YZ, XZ), (XY, YZ), (XY, Z), (X, Y, Z)\}$. We can express

$$\begin{aligned} G^2[(X, Y, Z)] = {} & G^2[(XY, YZ, XZ)] \\ & + \{G^2[(XY, YZ)] - G^2[(XY, YZ, XZ)]\} \\ & + \{G^2[(XY, Z)] - G^2[(XY, YZ)]\} \end{aligned}$$

$$
\begin{aligned}
&+ \{G^2[(X, Y, Z)] - G^2[(XY, Z)]\} \\
&= G^2[(XY, YZ, XZ)] \\
&\quad + G^2[(XY, YZ) \mid (XY, YZ, XZ)] \\
&\quad + G^2[(XY, Z) \mid (XY, YZ)] \\
&\quad + G^2[(X, Y, Z) \mid (XY, Z)]
\end{aligned}
$$

Thus we can first test whether model (XY, YZ, XZ) fits. Assuming that it does, the goodness of fit of (XY, YZ) can be tested through the second component. Assuming (XY, YZ) fits, we can then test the goodness of fit of (XY, Z). Finally, assuming (XY, Z) fits, the goodness of fit of (X, Y, Z) can be tested through the final component.

Suppose we consider the nested sequence $\{(DP, VP, DV), (VP, DV), (P, DV), (D, V, P)\}$ for the death penalty data. From Table 4.3, the four single degree of freedom components are 0.70, $1.88 - 0.70 = 1.18$, $8.13 - 1.88 = 6.25$, and $137.93 - 8.13 = 129.80$. We first accept the model (DP, VP, DV) and then accept (VP, DV), given (DP, VP, DV). At the third step the component 6.25 is large for $\text{df} = 1$, and we reject (P, DV). Hence the model (VP, DV) is the one selected.

The final model selected using this approach depends on the choice for the nested set of models. For example, the set $\{(DP, VP, DV), (DP, DV), (P, DV), (D, V, P)\}$ would have resulted in the choice of (DP, VP, DV). Before accepting a particular model using this conditional breakdown, it is a good idea to check also its goodness of fit according to the unconditional test. Each of several conditional components may be relatively large but not quite statistically significant, yet their sum may be significant.

Stepwise Procedures

In most applications the theory that guides the research suggests the analysis of certain models. In some cases, however, it may be of interest to use an automatic selection method analogous to forward selection and backward elimination procedures employed in multiple regression analysis (e.g., see Draper and Smith 1981, Ch. 6) to help select a model. Such an approach is simplified if models having all terms of various orders are first fitted. This helps to identify a restricted range of models for the search. For example, if the mutual independence model fits poorly, but the model having all two-factor associations fits well, we can then consider the intermediate models. In a forward selection approach, two-factor association terms are added sequentially to the mutual independence model until a good fit is achieved. At each stage the term is selected that gives the greatest improvement in fit. The maximum P-value for the resulting model is usually used as the criterion, since reductions in G^2 for different terms are based on different degrees of freedom. In a backward elimination procedure, two-factor associations are sequentially removed. At each stage the term removed is the one whose loss has the least damaging effect. For further details about these types of procedures, see Goodman (1971), Bishop et al. (1975, pp. 165–167), and Benedetti and Brown (1978).

Residuals

Once a preliminary model has been selected, it is useful to make a cell-by-cell comparison of the observed and estimated expected frequencies. The model may fit rather poorly in some of the cells, and this lack of fit may give us insight in explaining associations and interactions (e.g., Complement 4.3). The pattern of the deviations may even suggest an alternative model that would give a better fit. This is particularly true when some of the variables are ordinal, since the pattern of positive and negative deviations can indicate trends that are not well represented by the model.

A simple way of measuring cell deviations is through the *standardized residuals*, defined for cell u (in a table with arbitrary number of dimensions) by

$$R_u = \frac{(n_u - \hat{m}_u)}{\sqrt{\hat{m}_u}} \tag{4.3}$$

The squared standardized residuals are components of the Pearson chi-squared statistic, and $\sum R_u^2 = X^2$. If a certain model holds, these residuals are asymptotically normal with mean 0, and the average variance of the $\{R_u\}$ equals df/(number of cells). Since df $<$ number of cells, the asymptotic variance of R_u may be much less than 1.0 for a complex model. Hence the standardized residuals do not then behave as variably as standard normal deviates. Haberman (1973) has defined *adjusted residuals*, which are asymptotically standard normal. Unfortunately, the formula for the adjusted residual depends on the model. Haberman (1978, p. 275) gave a general expression for adjusted residuals for those models that have closed-form estimates of the expected frequencies. Adjusted residuals are calculated by the LOGLINEAR program in the computer package SPSS^X.

We now reconsider the dumping severity data of Table 2.4 to illustrate the use of residuals. In Section 2.3 we noted that there is slight evidence of an association between dumping severity and operation, as $X^2 = 10.54$ with df $= 6$. Table 4.7 lists the standardized residuals and the adjusted residuals for the independence model applied to these data. For this model Haber-

Table 4.7. Standardized Residuals (Adjusted Residuals in Parentheses) for the Independence Model Applied to the Dumping Severity Data

	Dumping Severity		
Operation	None	Slight	Moderate
A	0.77 (1.35)	−0.31 (−0.43)	−1.22 (−1.48)
B	1.05 (1.86)	−1.62 (−2.25)	0.30 (0.41)
C	−0.67 (−1.19)	1.02 (1.44)	−0.19 (−0.23)
D	−1.09 (−1.95)	0.85 (1.19)	1.04 (1.29)

Note: $\sum R^2 = X^2 = 10.5$, df = 6.

man's adjusted residuals have the form

$$R_{ij} = \frac{(n_{ij} - \hat{m}_{ij})}{\sqrt{\hat{m}_{ij}(1 - p_{i+})(1 - p_{+j})}}$$

where $\hat{m}_{ij} = n_{i+} n_{+j}/n$. Positive residuals occur where the level is high for both variables or low for both variables, whereas negative residuals occur where the level is high for one variable and low for the other. Thus the sample shows a tendency for greater severity of operation to be accompanied by more severe dumping than would be expected under independence. The next chapter presents a loglinear model for ordinal variables that represents well this tendency.

Effects of Sample Size

Researchers too rarely recognize the effects of sample size on the choice of a model through goodness-of-fit tests. A population association or interaction that is strong will likely be detected even if the sample size is small. However, very weak associations have a strong likelihood of being detected only with larger samples. As a consequence researchers having large sample sizes will typically need more complex models to pass goodness-of-fit tests than will researchers having small sample sizes.

Two cautionary remarks are suggested by these observations. Researchers having small data sets should realize that the true picture may be more complicated than suggested by the most parsimonious model they can accept in a goodness-of-fit test. On the other hand, researchers having very large sample sizes should realize that some associations or interactions that are statistically significant may be very weak and perhaps unimportant from a substantive viewpoint. Hence it may be possible to obtain a good representation of the data using a model that is simpler than (but has similar parameter estimates as) the simplest model that passes a goodness-of-fit test. Any analysis that focuses only on goodness-of-fit tests is therefore incomplete. We should also consider estimation of model parameters and description of the associations through measures such as the odds ratio in order to obtain a better understanding of the data.

In summary, researchers can gain much information from fitting various loglinear models to a cross-classification table. These models provide a logical framework for conducting tests of such hypotheses as independence, conditional independence, and no three-factor interaction. Through the estimates of model parameters and cell expected frequencies we can describe the nature of the associations and examine where lack of fit occurs. If an unsaturated model such as (XY, XZ, YZ) fits fairly well, its estimated cell probabilities can be regarded as smoothed values that dampen the random sampling fluctuations that occur in the sample proportions. In fact, if the true cell probabilities $\{\pi_{ijk}\}$ follow a pattern that is well approximated by a certain model, then the estimated cell probabilities $\{\hat{\pi}_{ijk}\}$ obtained using that model will tend to be better estimates of the $\{\pi_{ijk}\}$ than are the sample cell proportions $\{p_{ijk} = n_{ijk}/n\}$, unless the sample size is very large (see Bishop et al. 1975, pp. 313–315).

4.4. ITERATIVE PROPORTIONAL FITTING

We noted in Section 4.1 that there is no closed-form expression for ML estimates of expected frequencies for the model (XY, XZ, YZ). For tables with arbitrary numbers of dimensions, direct estimates of the $\{m_u\}$ do not exist in many cases, including those unsaturated models in which all two-factor associations are present. An iterative procedure can be used to obtain the $\{\hat{m}_u\}$ when direct estimates do not exist.

The *iterative proportional fitting* (IPF) algorithm, originally presented by Deming and Stephan (1940), is a simple method for obtaining the $\{\hat{m}_u\}$ for hierarchical loglinear models. The procedure may be outlined as follows:

1. Start with any initial estimates $\{\hat{m}_u^{(0)}\}$ whose association structure is no more complex than the model being fitted. For instance, $\{\hat{m}_u^{(0)} \equiv 1\}$ satisfies all the models we have discussed.

2. By multiplying the expected frequency estimates by appropriate scaling factors, successively adjust the $\{\hat{m}_u^{(0)}\}$ so that they match each marginal table in the set of sufficient statistics. For example, to fit the model (XY, XZ, YZ), the first cycle of the adjustment process has the following three steps:

$$\hat{m}_{ijk}^{(1)} = \hat{m}_{ijk}^{(0)}\left(\frac{n_{ij+}}{\hat{m}_{ij+}^{(0)}}\right)$$

$$\hat{m}_{ijk}^{(2)} = \hat{m}_{ijk}^{(1)}\left(\frac{n_{i+k}}{\hat{m}_{i+k}^{(1)}}\right)$$

$$\hat{m}_{ijk}^{(3)} = \hat{m}_{ijk}^{(2)}\left(\frac{n_{+jk}}{\hat{m}_{+jk}^{(2)}}\right)$$

Summing both sides of the first expression over k, we note that after the first step all $\hat{m}_{ij+}^{(1)} = n_{ij+}$, so that the observed and estimated expected frequencies match in the $X-Y$ marginal table. After the second step all $\hat{m}_{i+k}^{(2)} = n_{i+k}$, but the $X-Y$ marginal tables no longer match. After the third step all $\hat{m}_{+jk}^{(3)} = n_{+jk}$, but the $X-Y$ and $X-Z$ marginal tables no longer match. We begin a new cycle by rematching the $X-Y$ marginal tables through $\hat{m}_{ijk}^{(4)} = \hat{m}_{ijk}^{(3)}(n_{ij+}/\hat{m}_{ij+}^{(3)})$, and so forth.

3. As the cycles progress, the estimated expected frequencies tend to come simultaneously closer to matching the $\{n_{ijk}\}$ in the sufficient statistics, and they exhibit those associations represented in the model. The process is continued until the maximum difference between the observed and estimated frequencies in the marginal tables corresponding to the sufficient statistics is smaller than some set level.

Table 4.8 contains the result of the first full cycle (and more) of IPF for fitting the model (DP, VP, DV) to the death penalty data. After four cycles (12 steps), the $\{\hat{m}_{ijk}^{(12)}\}$ have two-way marginal frequencies that are all within 0.1 of the observed sufficient statistics. That is,

$$|\hat{m}_{ij+}^{(12)} - n_{ij+}| < 0.10$$

$$|\hat{m}_{i+k}^{(12)} - n_{i+k}| < 0.10$$

$$|\hat{m}_{+jk}^{(12)} - n_{+jk}| < 0.10$$

for all i, j, and k. The fourth-cycle estimates were the ones reported in Table 4.4, and they are also given in Table 4.8 for comparative purposes.

Birch (1963) showed that for hierarchical loglinear models there is only one set of cell estimates that satisfies the model and matches the observed data in the sufficient statistics, and that those estimates are the ML ones. Hence, if a solution $\{\hat{m}_u\}$ can be obtained that satisfies the model and

Table 4.8. Iterative Proportional Fitting of Model (DP, VP, DV) to Death Penalty Data

Cell			Fitted Marginal					
				DV	DP	VP	DV	DV
D	V	P	$\hat{m}^{(0)}$	$\hat{m}^{(1)}$	$\hat{m}^{(2)}$	$\hat{m}^{(3)}$	$\hat{m}^{(4)}$	$\hat{m}^{(12)}$
W	W	Y	1.0	75.5	17.93	22.06	22.03	18.67
		N	1.0	75.5	133.07	129.13	128.97	132.32
	B	Y	1.0	4.5	1.07	0.55	0.56	0.33
		N	1.0	4.5	7.93	8.38	8.44	8.68
B	W	Y	1.0	31.5	6.45	7.94	7.96	11.32
		N	1.0	31.5	56.55	54.87	55.04	51.70
	B	Y	1.0	51.5	10.55	5.45	5.44	5.68
		N	1.0	51.5	92.45	97.62	97.56	97.30

Note: W = White, B = Black, Y = yes, N = no.

matches the sufficient statistics, the $\{\hat{m}_u\}$ are necessarily the ML estimates. The reason the IPF algorithm works (i.e., gives ML estimates) is that it produces a sequence of estimates that converges to such a solution. For further details on uses of the IPF algorithm and guidelines about when direct estimates exist, see Bishop et al. (1975, pp. 76–102).

Other iterative methods can be used to obtain ML estimates in loglinear models. The Newton-Raphson method, described in Appendix B.1, is a general routine for finding the maximum of a function of several variables. For loglinear models it is applied to the log-likelihood function treated as a function of model parameters. Computations with the Newton-Raphson method are more complex than with iterative scaling routines, since matrix inversion is necessary in each cycle, but convergence usually occurs more rapidly.

4.5. LOGLINEAR MODELS AND ORDINAL INFORMATION

To illustrate further the fitting of loglinear models, let us consider the data in Table 4.9, a 4 × 3 × 4 cross-classification of operation, dumping severity, and hospital. The cross-classification in Table 2.4 is the two-dimensional marginal distribution of this table in which the results are combined for the four hospitals. Table 4.10 contains the results of fitting several loglinear models to the data, where O, D, and H symbolize the three variables. It is perhaps surprising to note that the complete independence model fits quite

Table 4.9. Cross-Classification of Operation, Hospital, and Dumping Severity

	Hospital											
	1			2			3			4		
	Dumping Severity											
Operation	N	S	M	N	S	M	N	S	M	N	S	M
A	23	7	2	18	6	1	8	6	3	12	9	1
B	23	10	5	18	6	2	12	4	4	15	3	2
C	20	13	5	13	13	2	11	6	2	14	8	3
D	24	10	6	9	15	2	7	7	4	13	6	4

Source: Grizzle, Starmer, and Koch (1969). Reprinted with permission from the Biometric Society.
Note: N = none, S = slight, M = moderate.

well. The value $G^2 = 32.61$ is less than df = 39 and has a P-value of $P = 0.76$ for testing goodness of fit.

Of the models containing an association term for only one pair of variables, (OD, H) fits best in terms of P-value and has $G^2 = 21.73$ based on df = 33. According to this model, dumping severity is associated with operation, but both these factors are independent of hospital. To construct a test of the hypothesis that D is independent of O, we form the test statistic

$$G^2[(O, D, H) \mid (OD, H)] = G^2[(O, D, H)] - G^2[(OD, H)]$$
$$= 32.61 - 21.73 = 10.88$$

Table 4.10. Goodness-of-Fit Tests for Loglinear Models Relating Operation (O), Dumping Severity (D), and Hospital (H)

Model	G^2	df	P-value
$(O, D, H)^a$	32.61	39	0.76
$(OD, H)^a$	21.73	33	0.93
$(HD, O)^a$	24.51	33	0.86
(OH, D)	31.64	30	0.38
$(HD, OD)^a$	13.63	27	0.98
(OD, OH)	20.76	24	0.65
(HD, OH)	23.54	24	0.49
(HD, OD, OH)	12.50	18	0.82
(OHD)	0	0	—

[a]These models are not appropriate if the H–O marginal table is considered fixed.

based on df = 39 − 33 = 6. This is a test of the null hypothesis that the λ^{OD} terms equal zero in the model (OD, H). This test has a P-value of about 0.10 and shows that there is some improvement in describing the data through (OD, H) instead of (O, D, H). Significant further improvement is not obtained by adding the O–H or D–H associations. Hence Table 4.10 suggests that the data are described adequately by (O, D, H) and (OD, H).

For simplicity, we considered models that apply for full multinomial sampling. In reality, the data seem to have been obtained by independent multinomial sampling within each O–H combination. In other words, the number of patients at each O–H combination should be treated as a fixed count. Because the sampling scheme fixes the O–H marginal table, only those models should be considered that contain the λ^{OH} term. For that set the simple models (D, OH) and (OD, OH) fit very well. The models are compared by testing that the λ^{OD} terms equal zero in the model (OD, OH). The statistic $G^2[(D, OH) \mid (OD, OH)] = 31.64 - 20.76 = 10.88$, based on df $= 30 - 24 = 6$, gives the same slight indication of O–D association as obtained previously.

Both statistics $G^2[(O, D, H) \mid (OD, H)]$ and $G^2[(D, OH) \mid (OD, OH)]$ for testing that all $\lambda^{OD} = 0$ equal the G^2 value of 10.88 obtained in Section 2.3 for testing independence in the marginal O–D table. This equivalence occurs because the O–D marginal association is identical to the O–D partial association, given H, for the models (OD, H) and (OD, OH).

The indication that (OD, OH) could be a better model than (D, OH) is supported by the $\{\hat{\lambda}^{OD}\}$ for the former model, the standardized values of which are reported in Table 4.11. Moderate positive $\hat{\lambda}^{OD}$ values occur in the corners of the table where operation severity and dumping severity both take high values or both take low values. Relatively large negative $\hat{\lambda}^{OD}$ values occur in the corners where one variable is at the high end and the other is at the low end of the scale. In other words, as the operation severity increases, there seems to be a tendency for dumping severity to increase.

Table 4.11. Standardized Values of O–D Association Parameter Estimates, for Model (OD, OH)

	Dumping Severity		
Operation	None	Slight	Moderate
A	1.59	0.53	−1.39
B	1.21	−2.00	0.80
C	−0.93	1.17	−0.23
D	−2.11	0.43	1.18

This tendency was also observed from the analysis of residuals in Table 4.7 for the independence model applied to the marginal O–D table.

It is very important to realize that the loglinear models discussed so far treat all the variables as nominal. Thus the $G^2[(D, OH) \mid (OD, OH)]$ statistic does not utilize the ordinal nature of dumping severity and of operation, and it cannot take advantage of the pattern in the $\hat{\lambda}^{OD}$ values observed in Table 4.11. Not surprisingly, therefore, we do not obtain as strong evidence of O–D association here as we will using ordinal methods in the remainder of the text. Specialized loglinear models that *do* treat O and D as ordinal are discussed in the next chapter.

NOTES

Section 4.1

1. The expressions for the minimal sufficient statistics can be obtained from the likelihood function. For full multinomial sampling, the kernel of the log-likelihood function has the form $\sum_i \sum_j \sum_k n_{ijk} \log m_{ijk}$. For the model symbolized by (XY, XZ, YZ), for instance, this simplifies to

$$n\mu + \sum_i n_{i++} \lambda_i^X + \sum_j n_{+j+} \lambda_j^Y + \sum_k n_{++k} \lambda_k^Z + \sum_i \sum_j n_{ij+} \lambda_{ij}^{XY} + \sum_i \sum_k n_{i+k} \lambda_{ik}^{XZ} + \sum_j \sum_k n_{+jk} \lambda_{jk}^{YZ}$$

   Since the multinomial distribution is in the exponential family, the sufficient statistics are the coefficients of the parameters. The minimal set in this case is $\{n_{ij+}\}$, $\{n_{i+k}\}$, and $\{n_{+jk}\}$.

2. There are alternative estimation methods that do give a closed-form expression for testing no three-factor interaction. For example, to test $\theta_{11(1)} = \theta_{11(2)}$ in a $2 \times 2 \times 2$ table we can use

$$z = \frac{(\log \hat{\theta}_{11(1)} - \log \hat{\theta}_{11(2)})}{\sqrt{\sum_i \sum_j \sum_k n_{ijk}^{-1}}}$$

   where $\hat{\theta}_{11(k)}$ is the sample odds ratio in table k [Recall formula (2.12)]. The square of this statistic is called the minimum logit chi-squared statistic for testing no three-factor interaction in a $2 \times 2 \times 2$ table. See Bishop et al. (1975, pp. 355–356).

3. Bishop et al. (1975, Ch. 12) discussed a Bayesian approach for estimating multinomial proportions. For a Dirichlet prior distribution

and a squared error loss function, the Bayes estimate of a proportion is a weighted average of the sample proportion and the prior mean, where the weight given to the sample increases as the sample size increases. Thus the Bayes estimate shrinks the sample data toward the prior mean. If the parameters $\{\beta_i\}$ for the Dirichlet prior distribution all equal the same constant $\beta > 0$, then the Bayes estimates are the same as the sample proportions calculated after β observations are added to each cell. Bishop et al. suggested "pseudo-Bayes" estimates in which the choice of $\sum \beta_i$ depends on the data. These biased estimates can have much smaller risk than the ML estimates when the number of cells is large, the sample size is small, and the true proportions are relatively near their prior means. See Efron and Morris (1977) for an interesting, elementary discussion of shrinkage estimation for proportions.

Section 4.2

4. The monotone decreasing behavior of G^2 as terms are added to a model is reminiscent of the behavior of the error sum of squares in regression analysis as explanatory variables are added to a model. This property is not shared by the Pearson statistic.

5. The *P*-values reported in tests such as the one comparing models (P, DV) and (VP, DV) are not strictly accurate, since the choice of models was based on viewing the results in Table 4.3 (rather than an a priori choice). Although the *P*-values do not strictly correspond to a probability of a more extreme result, they *do* order the models by a realistic criterion.

6. If the true cell probabilities satisfy a certain model, then the index G^2/df has an asymptotic mean of 1 as $n \to \infty$. Goodman (1971) suggested this index to compare the relative fits to a data set provided by several models. The larger the index, the poorer the fit.

7. The asymptotic joint distribution of model parameter estimates is multivariate normal. See Fienberg (1980, pp. 83–84) for an expression for the covariance structure.

Section 4.3

8. Brown (1976) suggested that in preliminary fitting of loglinear models, two conditional tests should be used to screen the importance of each possible term. In one test the term is the most complex parameter in a simple model, whereas in the other test all parameters of its level of complexity are included. For instance, to gauge whether λ^{XY} may be needed in a model for three variables, one conducts the marginal association test $G^2[(X, Y, Z) \mid (XY, Z)]$ and the partial association test $G^2[(XZ, YZ) \mid (XZ, YZ, XY)]$.

Section 4.4

9. The IPF algorithm can even be used to fit models having direct estimates. For such models the estimates are obtained at the end of the first cycle, except possibly when the number of dimensions exceeds six (see Haberman 1974a, p. 197).
10. For a set of cells S in a two-way table, two variables are *quasi independent* if we can express $\pi_{ij} = a_i b_j$ for (i, j) in S. One can fit this model by placing zeroes in cells not in S, placing ones in cells in S, and using IPF to successively fit the marginal frequencies of the cells in S. This model is often applied to square tables having identical row and column classifications, by letting S be the off-diagonal cells. See Goodman (1968) and Section 11.1 for further details.

COMPLEMENTS

1. Show that for the model (XZ, YZ), the Pearson or likelihood-ratio statistics can be expressed as $X^2 = \sum X_k^2$, where X_k^2 is the corresponding chi-squared statistic for testing independence between X and Y within level k of Z.
2. The data in Table 4.12 refer to whether an applicant was admitted into graduate school at the University of California at Berkeley, for the fall 1973 session. The data are presented by sex of applicant, for the six largest graduate departments at that university. Denote the three variables by A = whether admitted, S = sex, and D = department.
   a. Conduct a test of independence for the marginal A–S table. Using the odds ratio, interpret the result in terms of which sex has the higher admissions rate.
   b. The parenthesized values in the original table are estimated expected frequencies for the model (AD, AS, DS). Using these, compute the A–S partial odds ratio, and interpret it.
   c. Compare the A–S partial odds ratio from (b) to the marginal ob-

Table 4.12. Data for Complements 4.2 and 4.3

	Whether Admitted			
	Male		Female	
Department	Yes	No	Yes	No
A	512(529.3)	313(295.7)	89(71.7)	19(36.3)
B	353(353.6)	207(206.4)	17(16.4)	8(8.6)
C	120(109.3)	205(215.7)	202(212.8)	391(380.2)
D	138(137.2)	279(279.8)	131(131.8)	244(243.2)
E	53(45.7)	138(145.3)	94(101.3)	299(291.7)
F	22(23.0)	351(350.0)	24(23.0)	317(318.0)
Total	1198	1493	557	1278

Source: Data from Freedman et al. (1978, p. 14). See also Bickel et al. (1975).

served odds ratio in (a). Explain why they gave such different indications of the nature of the association between whether admitted and sex.

3. Refer to Complements 1 and 2 and Table 4.12. The results of fitting several loglinear models to the data are as follows:

Model	G^2
(A, D, S)	2097.7
(AD, S)	1242.3
(AS, D)	2004.2
(DS, A)	877.0
(AD, AS)	1148.9
(DS, AS)	783.6
(AD, DS)	21.7
(AD, AS, DS)	20.2
(ADS)	0.0

The next table gives the six single degree of freedom components of the G^2 statistic for the model (AD, DS). Note that the only lack of fit is for department A, and that the model fits very well for the other five departments ($G^2 = 2.7$, df $= 5$).

Department	G^2
A	19.05
B	0.26
C	0.75
D	0.30
E	0.99
F	0.38
Total	21.7

a. For department A, $\hat{m}_{111} = 531.4$ for the model (*AD*, *DS*). Interpret the nature of the lack of fit for that department, and explain how to interpret the data for the other five departments. How do your conclusions compare to the ones you made in Complement 2(c)?

b. The standardized residuals for the model (*AD*, *DS*) are given next. What would you conclude about the fit of this model based on a residual analysis?

		Admitted	
Department	Sex	Yes	No
A	Male	−0.84	1.13
	Female	2.33	−3.13
B	Male	−0.06	0.08
	Female	0.30	−0.39
C	Male	0.56	−0.41
	Female	−0.42	0.31
D	Male	−0.30	0.22
	Female	0.32	−0.23
E	Male	0.71	−0.41
	Female	−0.49	0.29
F	Male	−0.41	0.11
	Female	0.43	−0.11

c. The three-factor interaction model (*ADS*) is the only one that fits the entire $2 \times 2 \times 6$ table. How would you explain the nature of the three-factor interaction in this example?

4. Consider the sampling scheme for a three-way table in which the $\{n_{ijk}\}$

are regarded as independent Poisson random variables having unknown means $\{m_{ijk}\}$.

a. Let $z_{ijk} = (n_{ijk} - m_{ijk})/\sqrt{m_{ijk}}$. Show that the sum of variances of the $\{z_{ijk}\}$ equals the number of cells in the table.

b. Let $R_{ijk} = (n_{ijk} - \hat{m}_{ijk})/\sqrt{\hat{m}_{ijk}}$. If a model having certain expected frequencies $\{m_{ijk}\}$ holds, argue that the sum of variances of the $\{R_{ijk}\}$ asymptotically equals df for testing goodness of fit of the model. Hence, the average variance for individual R_{ijk}'s is less than 1.

5. Suppose that the row proportions $\{p_{i+}\}$ and the column proportions $\{p_{+j}\}$ are fixed for a $r \times c$ table.

a. Show how to use the IPF method to find a set of cell proportions $\{p_{ij}\}$ that have these marginal totals and for which the local odds ratios $\{\hat{\theta}_{ij} = p_{ij}p_{i+1,j+1}/p_{i,j+1}p_{i+1,j}\}$ all take on some fixed value $\hat{\theta}$. (Hint: Let the initial values be 1.0 in all cells in the first row and in all cells in the first column. This determines all other cell entries such that all local odds ratios equal $\hat{\theta}$. Use IPF to scale the cell entries to match successively row and column totals.)

b. More generally, show how to find $\{p_{ij}\}$ that have certain fixed values for row and column totals and have certain fixed values $\{\hat{\theta}_{ij}\}$ (not necessarily equal) for local odds ratios.

CHAPTER 5

Loglinear Models for Ordinal Variables

The loglinear models discussed in Chapter 4 have been widely used in recent years. They have gained popularity due to their exposure in books (e.g., the comprehensive one by Bishop et al. 1975) and due to the increasing availability of computer packages for fitting them. These loglinear models treat all variables as nominal, however, in the sense that parameter estimates and chi-squared statistics are invariant to orderings of categories. Thus these models fail to utilize all the available information when at least one of the variables is ordinal.

The analyses of the dumping severity data in Sections 4.3 and 4.5 illustrated problems that occur in treating ordinal variables as nominal. The residuals in Table 4.7 for the independence model showed lack of fit indicative of a positive association between dumping severity and operation severity. Similarly, the λ^{OD} estimates in Table 4.11 for a model containing O–D association parameters reflected this same trend in the association. Although the association was not significant, this nominal treatment of the data cannot take advantage of the monotonic manner in which the sample data depart from independence.

In this chapter loglinear models are described that *do* exploit the ordinal variables in the cross-classification. Among the advantages of using these models instead of the standard loglinear models of Chapter 4 are the following:

1. The association parameters in the models for ordinal variables describe certain types of trends and are easier to interpret than the general ones in the standard models.
2. There are unsaturated ordinal models for association in a two-way table and for three-factor interaction in a three-way table. The stan-

dard models are saturated for these cases. The ordinal models give a more structured form for the association and interaction terms, a form that does not require all the degrees of freedom.

3. Tests based on ordinal models have greater power for detecting certain important types of alternatives to the null hypotheses of independence, conditional independence, and no interaction.

The first section of this chapter considers a loglinear model for the two-way table in which both variables are ordinal, referred to as the ordinal-ordinal case. The second section considers a model for ordinal-nominal tables. The final two sections illustrate how these models can be generalized for multidimensional tables.

5.1. LOGLINEAR MODEL FOR ORDINAL-ORDINAL TABLES

For the two-dimensional table having expected frequencies $\{m_{ij}\}$, we rarely expect the independence model

$$\log m_{ij} = \mu + \lambda_i^X + \lambda_j^Y$$

to give an adequate fit. However, in the standard hierarchical system, the model of next greater complexity is the saturated one having an additional $(r-1)(c-1)$ independent λ_{ij}^{XY} parameters. Thus a nontrivial model does not exist for describing the association. If one or both variables are ordinal, though, simple models exist that are more complex and realistic than the independence model yet are unsaturated.

Suppose that both the column and row variables of the two-dimensional table are ordinal. We assume here that known scores $\{u_i\}$ and $\{v_j\}$ can be assigned to the rows and columns, respectively, where $u_1 < u_2 < \cdots < u_r$ and $v_1 < v_2 < \cdots < v_c$. In many applications the choice of scores will reflect assumed distances between midpoints of categories for an underlying interval scale. Equally spaced scores result in the simplest interpretation for the model discussed in this section. In practice, the integer scores $\{u_i = i\}$ and $\{v_j = j\}$ are most commonly used.

Uniform Association Model

A simple loglinear model that utilizes the orderings of the rows and the columns through these scores is given by

$$\log m_{ij} = \mu + \lambda_i^X + \lambda_j^Y + \beta(u_i - \bar{u})(v_j - \bar{v}) \tag{5.1}$$

where $\sum \lambda_i^X = \sum \lambda_j^Y = 0$. This model has only one more parameter (β) than the independence model, so it has df $= (r-1)(c-1) - 1 = rc - r - c$ for testing goodness of fit. Unlike the general model discussed in Section 2.6 for association between X and Y,

$$\log m_{ij} = \mu + \lambda_i^X + \lambda_j^Y + \lambda_{ij}^{XY} \tag{5.2}$$

model (5.1) does not require additional association parameters as the number of categories of X or Y increases. Model (5.1) is the special case of model (5.2) in which the general association term λ_{ij}^{XY} takes the structured form $\beta(u_i - \bar{u})(v_j - \bar{v})$.

The β parameter in model (5.1) describes the association between X and Y. The independence model is the special case in which $\beta = 0$. The association term $\beta(u_i - \bar{u})(v_j - \bar{v})$ reflects a deviation of $\log m_{ij}$ from the independence model. If $\beta > 0$, more observations are expected to have (large X, large Y) values or (small X, small Y) values than if X and Y are independent. If $\beta < 0$, on the other hand, the deviation of $\log m_{ij}$ from the independence model is positive for (large X, small Y) values and for (small X, large Y) values. In either case the deviation from independence increases in the directions of the four corner cells of the table.

In model (5.1) the deviation of $\log m_{ij}$ from the independence model is linear in Y for fixed X and linear in X for fixed Y. For a particular row i, for instance, the deviation is a linear function of Y (through the scores $\{v_j\}$) with slope $\beta(u_i - \bar{u})$. Because of this property, model (5.1) is sometimes referred to as the *linear-by-linear association model.* Similar models can be formulated that have nonlinear departures from independence (see Complement 5.8).

For the independence model the relative sizes of expected frequencies in two rows a and b, as measured by

$$\log\left(\frac{m_{bj}}{m_{aj}}\right) = \log m_{bj} - \log m_{aj}$$

is constant for all columns j. For model (5.1), however,

$$\log m_{bj} - \log m_{aj} = (\lambda_b^X - \lambda_a^X) + \beta(u_b - u_a)(v_j - \bar{v})$$

Thus the log odds of classification in row b instead of row a is a linear function of Y with slope $\beta(u_b - u_a)$. For any pair of rows $a < b$, there are relatively more large observations on Y in row b than in row a if $\beta > 0$. In fact the conditional distribution of Y in row b is stochastically higher than the conditional distribution of Y in row a if $\beta > 0$, and it is stochastically lower if $\beta < 0$.

The magnitude of β can be interpreted as follows. For an arbitrary pair of rows $a < b$ and an arbitrary pair of columns $c < d$,

$$\log\left(\frac{m_{ac}\,m_{bd}}{m_{ad}\,m_{bc}}\right) = \beta(u_b - u_a)(v_d - v_c) \tag{5.3}$$

In other words, the log odds ratio formed from a rectangular pattern of cells is directly proportional to the product of the distance between the rows and the distance between the columns. Hence β can be interpreted as the log odds ratio per unit distances $u_b - u_a = v_d - v_c = 1$ on X and Y; that is, the log odds ratio equals β whenever the rows are one unit apart and the columns are one unit apart.

Goodman (1979a) and Haberman (1974b) posed several loglinear models for ordinal variables. Goodman defined models in terms of the $(r-1)(c-1)$ local odds ratios

$$\theta_{ij} = \frac{(m_{ij}\,m_{i+1,\,j+1})}{(m_{i,\,j+1}\,m_{i+1,\,j})}$$

defined for adjacent rows and adjacent columns, $1 \leq i \leq r-1$ and $1 \leq j \leq c-1$. Goodman referred to the model in which all θ_{ij} are equal as the *uniform association model*. Such a model is the special case of model (5.1) in which the $\{u_i\}$ and $\{v_j\}$ are equal-interval scores; that is, $u_2 - u_1 = u_3 - u_2 = \cdots = u_r - u_{r-1}$ and $v_2 - v_1 = v_3 - v_2 = \cdots = v_c - v_{c-1}$. In particular, the integer scores $\{u_i = i\}$ and $\{v_j = j\}$ result in all $\theta_{ij} = \exp(\beta)$ and all $\log\theta_{ij} = \beta$. Unless an uneven spacing of the scores is natural, we suggest using these integer scores so that β can be interpreted simply as the common value of the local log odds ratio (see Figure 5.1).

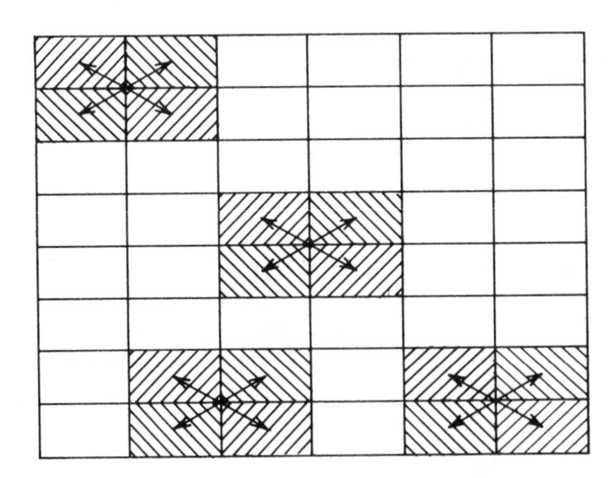

Figure 5.1. Constant odds ratio implied by uniform association model. (*Note:* β = the constant log odds ratio for adjacent rows and adjacent columns.)

Estimation of Model

Unfortunately there is no closed-form expression for the ML estimates $\{\hat{m}_{ij}\}$ of the $\{m_{ij}\}$ in model (5.1). Under the usual sampling assumptions, the estimates satisfy the likelihood equations

$$\hat{m}_{i+} = n_{i+}, \quad i = 1, \ldots, r$$

$$\hat{m}_{+j} = n_{+j}, \quad j = 1, \ldots, c$$

$$\sum_i \sum_j \hat{m}_{ij} u_i v_j = \sum_i \sum_j n_{ij} u_i v_j$$

The first two sets of equations indicate that the $\{\hat{m}_{ij}\}$ have the same row and column totals as the $\{n_{ij}\}$. Let $p_{ij} = n_{ij}/n$ and $\hat{\pi}_{ij} = \hat{m}_{ij}/n$ denote the estimates of the probability π_{ij} for the observed data and under the linear-by-linear association model, respectively. Then the third likelihood equation can be expressed as

$$\sum_i \sum_j u_i v_j \hat{\pi}_{ij} = \sum_i \sum_j u_i v_j p_{ij}$$

In other words, the expected value of XY is the same when the joint distribution is based on the estimated expected frequencies as it is when based on the observed data. Also all marginal expectations (e.g., EX, EY, EX^2, EY^2) are identical using either distribution since all $\hat{m}_{i+} = n_{i+}$ and $\hat{m}_{+j} = n_{+j}$. Hence the third constraint implies that the correlation between X and Y is the same using the fitted frequencies as it is using the observed data.

Goodman (1979a) gave an iterative solution for obtaining the $\{\hat{m}_{ij}\}$ using a unidimensional application of the Newton-Raphson method. The computer program ANOAS described in Appendix D.5 uses this method to fit the model. Nelder and Wedderburn (1972) and Haberman (1974b) used the general Newton-Raphson method, for which the ML estimates are the limit of a sequence of weighted least squares estimates. Convergence usually occurs very rapidly with this method. It is the basis of the computer package GLIM and the LOGLINEAR routine in the computer package SPSS^X, which can be used as shown in Appendixes D.1 and D.2 to fit all the loglinear models discussed in this chapter. Appendix D.4 shows how the computer package SAS (through its MATRIX procedure) can be used to fit ordinal loglinear models by the Newton-Raphson method. The Newton-Raphson methods are described in Appendix B.1.

An alternative method for fitting ordinal loglinear models, iterative scaling, follows from Corollary 2 to Theorem 1 of Darroch and Ratcliff (1972). This method is also discussed in Appendix B.1. For the linear-by-linear association model a single cycle has three steps. If $m_{ij}^{(t)}$ denotes the approximation for $\hat{m}_{ij}$ at a certain stage, then the three steps are (for all i and j)

$$m_{ij}^{(t+1)} = \left(\frac{n_{i+}}{m_{i+}^{(t)}}\right) m_{ij}^{(t)}$$

$$m_{ij}^{(t+2)} = \left(\frac{n_{+j}}{m_{+j}^{(t+1)}}\right) m_{ij}^{(t+1)}$$

$$m_{ij}^{(t+3)} = m_{ij}^{(t+2)} \left(\frac{\sum\sum u_a^* v_b^* n_{ab}}{\sum\sum u_a^* v_b^* m_{ab}^{(t+2)}}\right)^{u_i^* v_j^*} \left(\frac{\sum\sum (1 - u_a^* v_b^*) n_{ab}}{\sum\sum (1 - u_a^* v_b^*) m_{ab}^{(t+2)}}\right)^{1 - u_i^* v_j^*}$$

where the $\{u_i^*\}$ and $\{v_j^*\}$ are linear rescalings of the row scores and column scores that satisfy $0 \leq u_i^* \leq 1$ and $0 \leq v_j^* \leq 1$.

The initial estimates $\{m_{ij}^{(0)}\}$ in this routine can be taken equal to 1.0. The $\{m_{ij}^{(t+1)}\}$ satisfy $m_{i+}^{(t+1)} = n_{i+}$ for all i, and the $\{m_{ij}^{(t+2)}\}$ satisfy $m_{+j}^{(t+2)} = n_{+j}$ for all j. In fact the first two steps of the first cycle are the ones that the standard IPF method (discussed in Section 4.4) uses to fit the independence model, and give $m_{ij}^{(2)} = n_{i+} n_{+j}/n$. For the integer scores $\{u_i = i\}$ and $\{v_j = j\}$, in the third step we can take $u_i^* = (i-1)/(r-1)$ and $v_j^* = (j-1)/(c-1)$. Although this scaling method is relatively simple, the rate of convergence can be very slow. For large tables it sometimes takes several hundred cycles for adequate convergence.

Dumping Severity Example

Table 5.1 gives the estimated expected frequencies for fitting the independence model and the loglinear uniform association model to the cross-classification of dumping severity and operation. Notice how much better the uniform association model fits in the corners of the table, where it predicts the greatest departures from independence. The estimate of the association parameter is $\hat{\beta} = 0.163$ when integer scores are used in the model, and its standard error is 0.065. The positive value of $\hat{\beta}$ means that greater dumping severity tended to occur with greater operation severity in this sample. The estimated uniform odds ratio for adjacent rows and adjacent columns is $\hat{\theta}_{ij} = \exp(\hat{\beta}) = 1.18$. This value also equals the local odds ratios obtained using the estimated expected frequencies for the uniform association model; that is,

$$1.18 = \frac{(62.5 \times 30.9)}{(26.2 \times 62.9)} = \cdots = \frac{(35.3 \times 16.7)}{(13.7 \times 36.6)}$$

The goodness of fit of the two models is summarized in Table 5.2. As previously noted in Sections 2.3 and 4.5, the independence model (denoted by I) gives a barely adequate fit, with $G^2(I) = 10.88$ based on df = 6. The

Table 5.1. Observed Frequencies and Estimated Expected Frequencies for Independence Model and Loglinear Uniform Association Model for Dumping Severity Data

	Dumping Severity			
Operation	None	Slight	Moderate	Total
A	61	28	7	96
	(55.3)[a]	(29.7)	(11.0)	
	(62.5)[b]	(26.2)	(7.3)	
B	68	23	13	104
	(59.9)	(32.2)	(12.0)	
	(62.9)	(30.9)	(10.2)	
C	58	40	12	110
	(63.3)	(34.0)	(12.7)	
	(61.0)	(35.3)	(13.7)	
D	53	38	16	107
	(61.6)	(33.1)	(12.3)	
	(53.7)	(36.6)	(16.7)	

[a]For independence model $\hat{m}_{ij}$.
[b]For uniform association model $\hat{m}_{ij}$.

uniform association model (denoted by U) has one additional parameter and gives a very good fit, with $G^2(U) = 4.59$ based on df = 5.

Conditional Test of Independence

The statistical significance of the association between dumping severity and operation can be assessed by testing H_0: $\beta = 0$ in the uniform association model. This gives a conditional test of independence under the assumption that the uniform association model holds. The test statistic is the reduction

Table 5.2. Goodness-of-Fit of Models for Dumping Severity Data

Model	G^2	df
Independence	10.88	6
Uniform association	4.59	5
Independence, given uniform association	6.29	1

in G^2,

$$G^2(I \mid U) = G^2(I) - G^2(U) \tag{5.4}$$

with df = 1. For these data $G^2(I \mid U) = 10.88 - 4.59 = 6.29$ is the reduction in G^2 obtained by adding the association parameter β to the independence model. The P-value is about 0.01 for testing H_0: $\beta = 0$, and thus there is much stronger evidence of association than was obtained with the $G^2(I)$ statistic. If a positive association truly exists between dumping severity and operation, we have been more successful at detecting it using this ordinal approach.

If the uniform association model holds, the ordinal test based on $G^2(I \mid U)$ is asymptotically more powerful than the test based on $G^2(I)$ for detecting departures from independence. The $G^2(I)$ statistic cannot utilize as efficiently a constant pattern in local odds ratios, since it treats the variables as nominal.

Increased Power for Narrower Alternatives

When we have an expectation about the way variables are related, it is usually a good idea to use an alternative hypothesis and a test statistic that reflects it. The purpose in doing this is to obtain a test having strong power over a region of the parameter space that one feels is important. For instance, the $G^2(I \mid U)$ statistic has higher power than the $G^2(I)$ statistic for dependencies that are described by the uniform association model. The better the uniform association model approximates the true bivariate distribution, the more advantageous it is to use this ordinal conditional test.

An intuitive explanation can be given for the greater power of the $G^2(I \mid U)$ statistic. Suppose that the uniform association model truly holds, with $\beta \neq 0$. Then as the sample size increases, $G^2(U)$ stays relatively small, since it has a χ^2_{rc-r-c} distribution, whereas $G^2(I \mid U)$ grows in size and tends to be nearly as large as $G^2(I)$, since $G^2(I \mid U) = G^2(I) - G^2(U)$. The $G^2(I)$ statistic has df = $(r-1)(c-1)$, whereas the $G^2(I \mid U)$ statistic has df = 1. If $G^2(I \mid U)$ is nearly as large as $G^2(I)$, then it will yield a P-value smaller than does $G^2(I)$ because it is based on fewer degrees of freedom and hence will be relatively farther out in the tail of its null distribution. For example, for the dumping severity data, $G^2(I \mid U) = 6.29$ falls farther out in the tail of a χ^2_1 distribution ($P \cong 0.01$) than $G^2(I) = 10.88$ falls in the tail of a χ^2_6 distribution ($P \cong 0.10$).

In more theoretical terms, both the $G^2(I)$ and $G^2(I \mid U)$ statistics have asymptotic noncentral chi-squared distributions if $\beta \neq 0$. If the uniform association model holds, then the noncentrality for the $G^2(I \mid U)$ statistic is the same as the noncentrality for the $G^2(I)$ statistic, and the $G^2(I \mid U)$ statistic is more powerful because it is based on fewer degrees of freedom. This power comparison is based on a result of Das Gupta and Perlman (1974). Let $X^2_{\mathrm{df},\lambda}$ denote a noncentral chi-squared random variable with degrees of freedom df and noncentrality λ, and let $\chi^2_{\mathrm{df}}(\alpha)$ denote the $100(1-\alpha)$th percentile of a central chi-squared random variable with the same degrees of freedom. Das Gupta and Perlman showed that $P[X^2_{\mathrm{df},\lambda} > \chi^2_{\mathrm{df}}(\alpha)]$ increases as df decreases; that is, for fixed noncentrality, the probability of rejecting H_0 at a fixed α-level increases as the df of the test decreases. Therefore, in formulating chi-squared statistics, it is desirable to focus the anticipated noncentrality on a statistic having a small df value. See Fix et al. (1959) for a detailed discussion of these concepts.

One should not conclude that the ordinal test is uniformly more powerful than the $G^2(I)$ test. The $G^2(I)$ statistic, and the Pearson $X^2(I)$ statistic, will detect any deviation from independence, given enough data. The $G^2(I \mid U)$ statistic detects (with probability 1 as $n \rightarrow \infty$) only those deviations from independence for which the parameter that $\hat{\beta}$ estimates is nonzero. If in the population some local log odds ratios are positive and some are negative, for instance, then the uniform association model may fit poorly and the parameter that $\hat{\beta}$ estimates may equal zero (see Complement 5.1). In other words, independence implies that $\beta = 0$, but $\beta = 0$ does not imply independence if the uniform association model does not hold. Since the $G^2(I)$ statistic is designed to detect *any* departure from independence, it is better than the $G^2(I \mid U)$ statistic at detecting nonmonotonic dependencies for which β is zero or close to zero. The alternative hypothesis in the ordinal test ($\beta \neq 0$) is narrower than the broad alternative of "dependence," so the ordinal test sacrifices power for detecting dependencies that are not well summarized by the uniform association model. In many applications, though, the uniform association model (and more generally the linear-by-linear association model) describes the types of trends researchers envision as departures from independence for ordinal data.

Estimation

In most applications formal significance testing is of less interest than estimation of the strength of association. For large samples $\hat{\beta}$ is approximately normally distributed, and $\hat{\beta} \pm z_{p/2}\,\hat{\sigma}(\hat{\beta})$ is an approximate $100(1-p)$ percent confidence interval for β. The estimated asymptotic variance $\hat{\sigma}^2(\hat{\beta})$ is obtained from the inverse of the estimated information matrix. The covariance matrix of parameter estimates is automatically produced by the Newton-Raphson method for fitting loglinear models, as described in Appendix B.1.

The user can request this matrix with the computer packages GLIM and SPSS^X (LOGLINEAR).

The statistic $z = \hat{\beta}/\hat{\sigma}(\hat{\beta})$ provides an alternative way of testing H_0: $\beta = 0$. Unlike the $G^2(I \mid U)$ statistic, it retains information about direction of association, so it is useful for the alternatives H_a: $\beta > 0$ and H_a: $\beta < 0$.

5.2. LOGLINEAR MODEL FOR ORDINAL-NOMINAL TABLES

In many applications only one of the classifications in a two-way table is ordered. Even if both classifications are ordered, sometimes it is only relevant to use the ordinal nature of one of them (usually the response variable). In this section a loglinear model is discussed that treats the levels of X as nominal and the levels of Y as ordinal. We again use a set of scores $v_1 < v_2 < \cdots < v_c$, most commonly the integer scores, to reflect the ordering of the columns. As in the previous section our goal is to pose a model that is more complex than the independence model but not saturated. We do this by constructing an association term that has a linear departure of $\log m_{ij}$ from the independence model in the direction of the ordinal variable alone.

Row Effects Model

A loglinear model that utilizes the ordinal nature of Y through the scores $\{v_j\}$ is

$$\log m_{ij} = \mu + \lambda_i^X + \lambda_j^Y + \tau_i(v_j - \bar{v}) \tag{5.5}$$

where $\sum \lambda_i^X = \sum \lambda_j^Y = \sum \tau_i = 0$. Here the $\{v_j\}$ are known constants and the $\{\tau_i\}$ are parameters, $r - 1$ of which are linearly independent. Thus the model has

$$\text{df} = rc - [1 + (r-1) + (c-1) + (r-1)] = (r-1)(c-2)$$

and is unsaturated when $c > 2$.

The independence model is the special case of this model with all $\tau_i = 0$. The association term $\tau_i(v_j - \bar{v})$ in model (5.5) reflects the deviation of $\log m_{ij}$ from the independence model. The $\{\tau_i\}$ are row effects that can be interpreted as follows. Within a particular row the deviation of $\log m_{ij}$ from independence is a linear function of the ordinal variable, with slope τ_i. If $\tau_i > 0$, then in row i the probability of classification above $\bar{v}$ on Y is higher than would be expected if the variables were independent. If $\tau_i < 0$, observations in row i are more likely (compared to the independence case) to fall at the low end of the scale on the column variable Y.

For a fixed pair of rows a and b,

$$\log m_{bj} - \log m_{aj} = (\lambda_b^X - \lambda_a^X) + (\tau_b - \tau_a)(v_j - \bar{v})$$

That is, on a log scale the difference between proportions in two rows is a linear function of the ordinal variable Y, with slope the difference in the relevant τ parameters. Recall that this difference is a constant for the independence model. If $\tau_b > \tau_a$, the conditional Y distribution at the bth level of X is stochastically higher than the conditional Y distribution at the ath level of X. Hence the $\{\tau_i\}$ can be used to compare the rows in terms of how the responses on the ordinal variable tend to be distributed.

The $\{\tau_i\}$ can also be interpreted through odds ratios. For an arbitrary pair of rows a and b and an arbitrary pair of columns $c < d$,

$$\log\left(\frac{m_{ac}\, m_{bd}}{m_{ad}\, m_{bc}}\right) = (\tau_b - \tau_a)(v_d - v_c) \tag{5.6}$$

Hence this log odds ratio is proportional to the distance between the columns and is always positive if $\tau_b > \tau_a$. For the integer scores $\{v_j = j\}$, the log odds ratio is constant and equals $\tau_b - \tau_a$ for all $c - 1$ pairs of adjacent columns. Thus $\tau_b - \tau_a$ serves as a natural measure of the difference between two rows with respect to the conditional distribution of Y.

We will refer to model (5.5) as the *loglinear row effects model*. The corresponding *column effects model* has the association term $\rho_j(u_i - \bar{u})$. It treats the row variable as ordinal (with scores $\{u_i\}$) and the column variable as nominal. These models or special cases of them have been discussed by Simon (1974), Haberman (1974b), Goodman (1979a), Duncan (1979), Duncan and McRae (1979), and Fienberg (1980, pp. 61–64). Goodman's row effects model is the special case of model (5.5) in which the $\{v_j\}$ are integer scores.

The row effects model is naturally suited to the two-way table having nominal row variable and ordinal column variable, since the model has the same form (and produces the same G^2 value) if rows of the table are permuted. However, it can also be applied to ordinal-ordinal tables. This might be done if the departure of $\log m_{ij}$ from the independence model is not linear across the rows, as is necessary in model (5.1). It might also be done if the study is mainly concerned with comparing levels of the row variable with respect to their conditional distribution on the column variable, which is an ordinal response variable. The linear-by-linear association model (5.1) is the special case of the row effects model in which $\tau_i = \beta(u_i - \bar{u})$.

Estimation of Model

The ML estimates of expected frequencies for the row effects model satisfy the likelihood equations

$$\hat{m}_{i+} = n_{i+}, \qquad 1 \leq i \leq r$$

$$\hat{m}_{+j} = n_{+j}, \qquad 1 \leq j \leq c$$

$$\sum_j \hat{m}_{ij} v_j = \sum_j n_{ij} v_j, \qquad 1 \leq i \leq r$$

The third equation implies that for each row, the mean of the conditional distribution across the columns is the same when the conditional distribution is based on the estimated expected frequencies as it is when based on the observed data (see Complement 5.5).

Like the linear-by-linear association model, the row effects model does not have a closed-form expression for the ML estimates of expected frequencies. The estimates can be obtained using the Newton-Raphson method outlined in Appendix B.1. The computer packages GLIM and SPSS^X (LOGLINEAR routine) use this method to fit the model, as described in Appendixes D.1 and D.2. The Newton-Raphson method for fitting this model can also be easily programmed using the procedure MATRIX in the computer package SAS, as illustrated in Appendix D.4.

Other methods can be used to obtain the ML estimates. Goodman (1979a) described how to fit the model using the unidimensional version of the Newton-Raphson method. That approach is also described in Appendix B.1. The computer program ANOAS described in Appendix D.5 uses it to fit the row effects model. Simon (1974) and Fienberg (1980, p. 63) used the iterative scaling procedure suggested by Darroch and Ratcliff (1972). A single cycle has the three steps

$$m_{ij}^{(t+1)} = m_{ij}^{(t)}\left(\frac{n_{i+}}{m_{i+}^{(t)}}\right)$$

$$m_{ij}^{(t+2)} = m_{ij}^{(t+1)}\left(\frac{n_{+j}}{m_{+j}^{(t+1)}}\right)$$

$$m_{ij}^{(t+3)} = m_{ij}^{(t+2)}\left(\frac{\sum_b v_b^* n_{ib}}{\sum_b v_b^* m_{ib}^{(t+2)}}\right)^{v_j^*}\left(\frac{\sum_b (1 - v_b^*) n_{ib}}{\sum_b (1 - v_b^*) m_{ib}^{(t+2)}}\right)^{1 - v_j^*}$$

where the $\{v_j^*\}$ are a linear rescaling of the column scores $\{v_j\}$ that satisfy $0 \leq v_j^* \leq 1$. The first two steps in a cycle are the same as the ones for the linear-by-linear association model.

Table 5.3, based on data presented by Hedlund (1978), presents the re-

Table 5.3. Observed Frequencies and Estimated Expected Frequencies for Independence Model and for Loglinear Row Effects Model for Political Ideology Data

	Political Ideology			
Party Affiliation	Liberal	Moderate	Conservative	Total
Democrat	143 (102.0)[a] (136.6)[b]	156 (161.4) (168.7)	100 (135.6) (93.6)	399
Independent	119 (120.2) (123.8)	210 (190.1) (200.4)	141 (159.7) (145.8)	470
Republican	15 (54.7) (16.6)	72 (86.6) (68.9)	127 (72.7) (128.6)	214

[a]For independence model $\hat{m}_{ij}$.
[b]For row effects model $\hat{m}_{ij}$.

lationship between the ordinal variable "political ideology" and the nominal variable "political party affiliation" for a sample of voters taken in the 1976 presidential primary in Wisconsin. Table 5.3 contains estimated expected frequencies for the independence model and for the row effects model with integer scores applied to these data. The row effects model seems to give a much better fit, particularly at the ends of the ordinal scale where the greatest deviation from independence is predicted by this model. The estimated effects of party affiliation on ideology are $\hat{\tau}_1 = -0.495$, $\hat{\tau}_2 = -0.224$, and $\hat{\tau}_3 = 0.719$. The further $\hat{\tau}_i$ falls in the positive direction, the greater the tendency for the ith party to locate at the conservative end of the ideology scale. Hence in this sample the Democrats (row 1) are the least conservative group, and the Republicans (row 3) are much more conservative than the other two groups.

The row effects model with integer scores predicts constant odds ratios for adjacent columns of political ideology. To illustrate, $\hat{\tau}_3 - \hat{\tau}_1 = 1.214$ means that the odds of being classified conservative instead of moderate and the odds of being classified moderate instead of liberal are $\exp(1.214) = 3.37$ times higher for Republicans than for Democrats. These odds ratios can also be obtained from the estimated expected frequencies for the row effects model; for example,

$$3.37 = \frac{(168.7 \times 128.6)}{(93.6 \times 68.9)} = \frac{(136.6 \times 68.9)}{(168.7 \times 16.6)}$$

Table 5.4. Goodness of Fit of Models for Political Ideology Data

Model	G^2	df
Independence	105.66	4
Row effects	2.81	2
Independence, given row effects	102.85	2

Similarly, $\exp(\hat{\tau}_2 - \hat{\tau}_1) = 1.31$ and $\exp(\hat{\tau}_3 - \hat{\tau}_2) = 2.57$ are the constant odds ratios for comparing Independents to Democrats and for comparing Republicans to Independents, respectively.

The goodness of fit of the independence model (denoted by I) and of the row effects model (denoted by R) is summarized in Table 5.4. The independence model is completely inadequate, with $G^2 = 105.66$ based on $\text{df} = (r-1)(c-1) = 4$. The addition of the two linearly independent row effect parameters results in a much better fit, with $G^2(R) = 2.81$ based on $\text{df} = (r-1)(c-2) = 2$.

Conditional Test of Independence

We can assess the statistical significance of the association between the two variables in the row effects model by testing H_0: $\tau_1 = \cdots = \tau_r = 0$. If model (5.5) holds, this homogeneity of the row effects corresponds to independence. A conditional test of independence can therefore be based on

$$G^2(I \mid R) = G^2(I) - G^2(R) \tag{5.7}$$

with $\text{df} = (r-1)(c-1) - (r-1)(c-2) = r - 1$. Given that the row effects model holds, this test is asymptotically more powerful at detecting an association than the $G^2(I)$ test, since it concentrates the noncentrality on fewer degrees of freedom. The $G^2(I)$ test ignores the ordinal nature of Y, whereas the $G^2(I \mid R)$ test focuses on alternatives where the ordinal scaling is utilized through a linear departure of $\log m_{ij}$ from independence. These alternatives are narrower and usually of greater interest than the general alternative to H_0: independence. For the political ideology data, $G^2(I \mid R) = 105.66 - 2.81 = 102.85$ based on $\text{df} = 2$ shows very strong evidence of an association.

To compare a particular pair of rows, one can test H_0: $\tau_a = \tau_b$ by applying $G^2(I \mid R)$ to the $2 \times c$ table that those rows form. This test has $\text{df} = 1$. Alternatively, one can form a confidence interval for $\tau_b - \tau_a$ (see Note 5.3). For Table 5.3 these analyses indicate strong evidence of a difference between each pair of party affiliations and give the conclusion $\tau_1 < \tau_2 < \tau_3$.

5.3. ORDINAL LOGLINEAR MODELS FOR HIGHER DIMENSIONS

The models in Sections 5.1 and 5.2 can be readily generalized to multidimensional tables having at least one ordinal variable. We will illustrate for the $r \times c \times l$ cross-classification of three variables X, Y, and Z having expected frequencies $\{m_{ijk}\}$.

The hierarchical models of usual interest for three dimensions range from the simple mutual independence model

$$\log m_{ijk} = \mu + \lambda_i^X + \lambda_j^Y + \lambda_k^Z \tag{5.8}$$

to the model that contains all the partial association terms but no three-factor interaction term,

$$\log m_{ijk} = \mu + \lambda_i^X + \lambda_j^Y + \lambda_k^Z + \lambda_{ij}^{XY} + \lambda_{ik}^{XZ} + \lambda_{jk}^{YZ} \tag{5.9}$$

The next most complex model beyond (5.9) has the $(r-1)(c-1)(l-1)$ independent three-factor interaction parameters $\{\lambda_{ijk}^{XYZ}\}$ and is of little interest because it is saturated. If one or more of the variables in the cross-classification are ordinal, though, there is a richer hierarchy of models that includes:

1. Partial association models that are more parsimonious and simpler to interpret than model (5.9).
2. Three-factor interaction models that are unsaturated and also easily interpretable.

In interpreting these models, we refer to the odds ratios $\{\theta_{ij(k)}\}$, $\{\theta_{i(j)k}\}$, and $\{\theta_{(i)jk}\}$ that describe the local conditional associations between two variables within a fixed level of the third variable [see (3.1)]. The ratio of odds ratios

$$\theta_{ijk} = \frac{\theta_{ij(k+1)}}{\theta_{ij(k)}} = \frac{\theta_{i(j+1)k}}{\theta_{i(j)k}} = \frac{\theta_{(i+1)jk}}{\theta_{(i)jk}} \tag{5.10}$$

is used for describing local three-factor interaction. In other words, θ_{ijk} describes the interaction in a $2 \times 2 \times 2$ section of the table consisting of adjacent rows, adjacent columns, and adjacent layers. There is an absence of three-factor interaction if all $(r-1)(c-1)(l-1)$ of the θ_{ijk} equal 1.0. This section considers models that assume an absence of three-factor interaction. Models that permit interaction are discussed in Section 5.4.

All Variables Ordinal

We first consider the case in which X, Y, and Z are all ordinal. Let $\{u_i\}$, $\{v_j\}$, and $\{w_k\}$ represent scores assigned to the levels of X, Y, and Z, respectively. A simple generalization of the mutual independence model that has an association term for each pair of variables but utilizes their ordinal nature is

$$\log m_{ijk} = \mu + \lambda_i^X + \lambda_j^Y + \lambda_k^Z + \beta^{XY}(u_i - \bar{u})(v_j - \bar{v}) + \beta^{XZ}(u_i - \bar{u})(w_k - \bar{w}) + \beta^{YZ}(v_j - \bar{v})(w_k - \bar{w}) \qquad (5.11)$$

The β^{XY}, β^{XZ}, and β^{YZ} parameters describe the pairwise partial associations. This model has only three more parameters than the independence model, so df $= rcl - r - c - l - 1$, and the model is always unsaturated.

In model (5.11) the association term for each pair of variables has the same form as the association term in the linear-by-linear association model (5.1). The departure of $\log m_{ijk}$ from the independence model is a linear function of X for fixed Y and Z, a linear function of Y for fixed X and Z, and a linear function of Z for fixed X and Y. There is no three-factor interaction term in the model, and it is easily seen that all θ_{ijk} equal 1.0. Hence the partial association between each pair of variables is the same for all levels of the third variable.

It is simplest to apply model (5.11) with integer scores for all three variables. In that case the local conditional log odds ratios simplify to

$$\begin{aligned} \log \theta_{ij(k)} &= \beta^{XY} \\ \log \theta_{i(j)k} &= \beta^{XZ} \\ \log \theta_{(i)jk} &= \beta^{YZ} \end{aligned} \qquad (5.12)$$

According to (5.12), the local odds ratio is uniform for each pair of variables, and the strength of association is homogeneous across the levels of the third variable. Model (5.11) can therefore be described as a *homogeneous uniform association model.*

If this model fits well, it is very easy to describe the patterns of association in the table. If this model holds, and if a particular β equals zero, then there is conditional independence between those variables. A test of conditional independence of X and Y (given Z) with df $= 1$ can be based on the difference between G^2 for the model with $\beta^{XY} = 0$ and G^2 for model (5.11).

In many applications model (5.11) is too simple to give a good fit. In some cases it is not even relevant to utilize the orderings of classifications for all the ordinal variables. More complex association terms can then be used for some of the pairs of variables. For instance, models discussed in the remainder of this section can be used even if all variables are ordinal, when the ordinal nature of some variables is not relevant to the analysis, when departures from independence are not linear, or when the sampling design dictates fitting certain marginal distributions.

Ordinal and Nominal Variables

The row effects model for the nominal-ordinal table can be generalized for the multidimensional table consisting of some ordinal and some nominal variables. For example, if X is nominal and Y and Z are ordinal, a basic model is

$$\log m_{ijk} = \mu + \lambda_i^X + \lambda_j^Y + \lambda_k^Z + \tau_i^{XY}(v_j - \bar{v}) + \tau_i^{XZ}(w_k - \bar{w}) + \beta^{YZ}(v_j - \bar{v})(w_k - \bar{w}) \tag{5.13}$$

where $\sum \lambda_i^X = \sum \lambda_j^Y = \sum \lambda_k^Z = \sum \tau_i^{XY} = \sum \tau_i^{XZ} = 0$. For this case this is usually the simplest model that would be formulated to include all pairwise associations but no three-factor interaction. The Y–Z association term for the ordinal variables has the same form as the association term for the linear-by-linear association model (5.1). The X–Y and X–Z association terms for the pairs of nominal and ordinal variables have the same form as the association term for the row effects model (5.5).

For model (5.13) applied with integer scores for the ordinal variables,

$$\begin{aligned} \log \theta_{ij(k)} &= \tau_{i+1}^{XY} - \tau_i^{XY} \\ \log \theta_{i(j)k} &= \tau_{i+1}^{XZ} - \tau_i^{XZ} \\ \log \theta_{(i)jk} &= \beta^{YZ} \\ \log \theta_{ijk} &= 0 \end{aligned} \tag{5.14}$$

Here β pertains to the uniform association between Y and Z, which is homogeneous across levels of X. The $\{\tau_i^{XY}\}$ represent row effects of X on the X–Y association that are homogeneous across levels of Z. Thus the ordering of their values indicates how the levels of X are stochastically ordered with respect to the conditional Y distributions (within each level of Z). Similarly, the $\{\tau_i^{XZ}\}$ represent row effects of Z on the X–Z association that are homogeneous across levels of Y.

Suppose now that two of the three variables are nominal, say X and Y. A basic model is then

$$\log m_{ijk} = \mu + \lambda_i^X + \lambda_j^Y + \lambda_k^Z + \lambda_{ij}^{XY} + \tau_i^{XZ}(w_k - \bar{w}) + \tau_j^{YZ}(w_k - \bar{w}) \tag{5.15}$$

where $\sum \lambda_i^X = \sum \lambda_j^Y = \sum \lambda_k^Z = \sum \tau_i^{XZ} = \sum \tau_j^{YZ} = \sum_i \lambda_{ij}^{XY} = \sum_j \lambda_{ij}^{XY} = 0$. A general association term $\{\lambda_{ij}^{XY}\}$ is used for the association between the nominal variables. The orderings of the $\{\tau_i^{XZ}\}$ and $\{\tau_j^{YZ}\}$ parameters indicate

Table 5.5. Association Terms and df for Loglinear Models in Three Dimensions with Linear Ordinal Effects

Ordinal Variables	Association Terms			df
	X–Y	X–Z	Y–Z	
X, Y, Z	$\beta^{XY}(u_i - \bar{u})(v_j - \bar{v})$	$\beta^{XZ}(u_i - \bar{u})(w_k - \bar{w})$	$\beta^{YZ}(v_j - \bar{v})(w_k - \bar{w})$	$rcl - r - c - l - 1$
Y, Z	$\tau_i^{XY}(v_j - \bar{v})$	$\tau_i^{XZ}(w_k - \bar{w})$	$\beta^{YZ}(v_j - \bar{v})(w_k - \bar{w})$	$rcl - 3r - c - l + 3$
Z	λ_{ij}^{XY}	$\tau_i^{XZ}(w_k - \bar{w})$	$\tau_j^{YZ}(w_k - \bar{w})$	$rcl - rc - r - c - l + 3$

how the levels of X and the levels of Y are stochastically ordered with respect to the conditional distributions on the ordinal variable Z.

Table 5.5 summarizes the models considered in this section and lists their degrees of freedom for testing goodness of fit. For $2 \times 2 \times 2$ tables these models are equivalent to the standard model (5.9). For larger tables, though, they are more parsimonious and simpler to interpret because of the structured associations. All three models incorporate the ordinal variables in the association terms through linear departures from conditional independence.

The models discussed in this section can be easily fitted using the computer packages GLIM and SPSS^X (LOGLINEAR routine). More complex models of this nature can be constructed by permitting nonlinear deviations from independence in the ordinal variables or three-factor interaction. Also, if these models do not fit well but the more general model (5.9) does, there may be an intermediate model that fits well and that is simpler than (5.9). For instance, suppose the X–Z and Y–Z association terms in (5.11) are replaced by the general terms λ_{ik}^{XZ} and λ_{jk}^{YZ}. We might use λ_{ik}^{XZ} for the X–Z association term, even though X and/or Z are ordinal, if the X–Z marginal table is fixed by the sampling design. The resulting model is still simpler to interpret than model (5.9), since it displays homogeneous linear-by-linear X–Y association.

Dumping Severity Example

In Section 4.5 we saw that the loglinear model (D, OH) fits the three-dimensional version of the dumping severity data (Table 4.9) fairly well, with $G^2 = 31.64$ based on df $= 30$. Some improvement was provided by the model (OD, OH), for which $G^2 = 20.76$ based on df $= 24$. However, these models did not exploit the category orderings for D and O. A simpler model than (OD, OH) that also assumes O–D partial association but has greater df from incorporating their ordinal nature is

$$\log m_{ijk} = \mu + \lambda_i^O + \lambda_j^H + \lambda_k^D + \lambda_{ij}^{OH} + \beta^{OD}(u_i - \bar{u})(w_k - \bar{w}) \qquad (5.16)$$

When this model is fitted using integer scores, $G^2 = 25.35$ based on df $= 29$, a very good fit. According to this model, dumping severity and operation have a uniform association that is the same for each hospital, and dumping severity is independent of hospital for each operation. The general association term λ_{ij}^{OH} is used for the O–H association, rather than the term $\tau_j^{OH}(u_i - \bar{u})$, because the O–H marginal distribution is regarded as fixed for these data.

Given that model (5.16) fits, the conditional independence of dumping severity and operation corresponds to $\beta^{OD} = 0$. The test statistic for testing H_0: $\beta^{OD} = 0$ is the difference of 6.29 between $G^2[(D, OH)] = 31.64$ and

$G^2 = 25.35$ for model (5.16), based on df = 1. Although the model (D, OH) whereby dumping severity is jointly independent of operation and hospital fits well, it gives a significantly poorer fit than the O–D uniform association model. In addition model (5.16) does not fit more poorly than the model (OD, OH) having a general O–D association term, as $G^2 = 25.35 - 20.76 = 4.59$ is based on df $= 29 - 24 = 5$. It also does not give a poorer fit than models having hospital effects on dumping severity. Table 5.6 summarizes the results of fitting several nested loglinear models to Table 4.9. In summary, model (5.16) adequately describes the data and also provides stronger evidence of an O–D association than is obtained by treating all variables as nominal.

The ML estimate of the O–D association parameter in model (5.16) is $\hat{\beta}^{OD} = 0.163$. This means that the estimated odds of moderate instead of slight dumping, or of slight instead of no dumping, are exp (0.163) = 1.18 times higher for each additional 25 percent of stomach removal. Model (5.16) implies that Table 4.9 can be collapsed over the H variable to study the O–D association, since H and D are conditionally independent in that model. Hence the estimate of β^{OD} in model (5.16) and the result of the conditional test of independence are the same as were obtained in Section 5.1 for the uniform association model applied to the marginal O–D table.

5.4. INTERACTION MODELS

There are many ways that the models introduced in Section 5.3 can be generalized to include interaction terms. A simple model for the case when X, Y, and Z are ordinal is

$$\log m_{ijk} = \mu + \lambda_i^X + \lambda_j^Y + \lambda_k^Z + \beta^{XY}(u_i - \bar{u})(v_j - \bar{v}) + \beta^{XZ}(u_i - \bar{u})(w_k - \bar{w}) + \beta^{YZ}(v_j - \bar{v})(w_k - \bar{w}) + \beta^{XYZ}(u_i - \bar{u})(v_j - \bar{v})(w_k - \bar{w}) \qquad (5.17)$$

Table 5.6. Analysis of Loglinear Models for Dumping Severity Data

Association Terms						
O–H	O–D	H–D	G^2	df	Difference in G^2	Difference in df
λ_{ij}^{OH}	—	—	31.64	30		
					6.29	1
λ_{ij}^{OH}	$ik\beta^{OD}$	—	25.35	29		
					4.59	5
λ_{ij}^{OH}	λ_{ik}^{OD}	—	20.76	24		
					8.26	6
λ_{ij}^{OH}	λ_{ik}^{OD}	λ_{jk}^{HD}	12.50	18		

This model has only one more parameter than model (5.11), so df $= rcl - r - c - l - 2$, and the model is unsaturated whenever r, c, or l exceeds two.

Uniform Interaction

When model (5.17) is applied with integer scores,

$$\log \theta_{ijk} = \beta^{XYZ}$$
$$\log \theta_{ij(k)} = \beta^{XY} + \beta^{XYZ}\left[k - \frac{(l+1)}{2}\right] \tag{5.18}$$

Hence the local interaction is constant and equals β^{XYZ} for all $2 \times 2 \times 2$ subtables formed from adjacent rows, adjacent columns, and adjacent layers. Model (5.17) can then be described as a *uniform interaction model.* Note from (5.18) that within a particular level of Z, the association between X and Y is uniform with local log odds ratio $\beta^{XY} + \beta^{XYZ}[k - (l+1)/2]$. Thus the strength of X–Y partial association is constant within levels of Z but changes linearly across the levels of Z. Similar remarks apply to the X–Z and Y–Z partial associations. If $\beta^{XYZ} = 0$ there is no three-factor interaction and model (5.17) simplifies to model (5.11).

More general interaction models have less structured associations for some or all pairs of variables. For example, suppose that the interaction term from model (5.17) is added to the standard loglinear model (XY, XZ, YZ), yielding

$$\log m_{ijk} = \mu + \lambda_i^X + \lambda_j^Y + \lambda_k^Z + \lambda_{ij}^{XY} + \lambda_{ik}^{XZ} + \lambda_{jk}^{YZ}$$
$$+ \beta^{XYZ}(u_i - \bar{u})(v_j - \bar{v})(w_k - \bar{w}) \tag{5.19}$$

This model has df $= (r-1)(c-1)(l-1) - 1$, so unlike the interaction model (XYZ) for nominal variables, it is unsaturated whenever r, c, or l exceeds two. There is a constant value for $\log \theta_{ijk}$ when this model is applied with integer scores, and that form of it is the most general uniform interaction model. Due to the general forms for the association terms, this model does not have uniform association within the partial tables.

Model (5.19) with $\beta^{XYZ} = 0$ corresponds to model (XY, XZ, YZ), in which there is an absence of three-factor interaction. Given that model (5.19) holds, the difference in G^2 values between model (XY, XZ, YZ) and model (5.19) gives a single degree of freedom test of no three-factor interaction (H_0: $\beta^{XYZ} = 0$).

Heterogeneous Uniform Association

Model (5.17) applied with integer scores may be described as a type of *heterogeneous uniform association* model, since the strength of the uniform partial association for each pair of variables changes across the levels of the third variable. We now consider a more general example of a heterogeneous uniform association model. We may wish to regard one of the variables (say Z) as a control variable, and assume a uniform conditional association between X and Y that may change in an unspecified manner across the levels of Z. Fitting such a model corresponds to fitting the bivariate uniform association model to X and Y separately within each level of Z. A summary measure of goodness of fit is obtained by summing the chi-squared values corresponding to the separate fits and summing their degrees of freedom. In this approach the interaction is not uniform, and the variable Z is treated as nominal. This model contains (5.17) as a special case, and it has $\text{df} = l(rc - r - c)$. For general scores it can be expressed as

$$\log m_{ijk} = \mu + \lambda_i^X + \lambda_j^Y + \lambda_k^Z + \lambda_{ik}^{XZ} + \lambda_{jk}^{YZ} + \beta_k^{XY}(u_i - \bar{u})(v_j - \bar{v}) \qquad (5.20)$$

For model (5.20) with integer scores, β_k^{XY} is the constant value of the X–Y local log odds ratio at level k of Z. That model is the most general form of heterogeneous X–Y uniform association model. Unlike model (5.17), model (5.20) does not also assume uniform X–Z and Y–Z partial associations.

Sections 5.3 and 5.4 have given a brief sampling of the types of models that can be fitted to multidimensional tables having ordinal classifications. For further discussions of these and other models, see Goodman (1981c), Clogg (1982b), and Agresti and Kezouh (1983).

Effect of Smoking Example

Table 5.7, taken from Forthofer and Lehnen (1981, p. 21), concerns associations among smoking status (S), breathing test results (B), and age (A) for Caucasians in certain industrial plants in Houston, Texas, during 1974 to 1975. The standard partial association model (SA, SB, BA) fits poorly, with $G^2 = 25.9$ based on $\text{df} = 4$, so we consider models having three-factor interaction terms.

The uniform interaction (integer scores) version of model (5.19) has only one additional parameter but fits strikingly better, as $G^2 = 2.7$ based on $\text{df} = 3$. The difference $25.9 - 2.7 = 23.2$ based on $\text{df} = 4 - 3 = 1$ provides extremely strong evidence that the three-factor interaction parameter in model (5.19) is nonzero. In other words, there is strong evidence that the association between smoking status and breathing test results depends on age. The ML estimate of the uniform local interaction measure $\log \theta_{ijk}$ is

Table 5.7. Cross-Classification of Houston Industrial Workers by Breathing Test Results, Smoking Status, and Age

		Breathing Test Results		
Age	Smoking Status	Normal	Borderline	Abnormal
<40	Never Smoked	577	27	7
	Former Smoker	192	20	3
	Current Smoker	682	46	11
40–59	Never Smoked	164	4	0
	Former Smoker	145	15	7
	Current Smoker	245	47	27

Source: *Public Program Analysis*, by R. N. Forthofer and R. G. Lehnen. © 1981 by Lifetime Learning Publications, Belmont, California 94002, a division of Wadsworth, Inc. Reprinted by permission of the publisher.

$\hat{\beta}^{SBA} = 0.83$, with standard error of 0.19. The association between smoking status and breathing test result is more positive at the higher age level. Any local odds ratio for the higher age group is estimated to be $\hat{\theta}_{ij1} = \hat{\theta}_{ij(2)}/\hat{\theta}_{ij(1)} = \exp(0.83) = 2.3$ times higher than the corresponding local odds ratio for the lower age group.

The heterogeneous uniform S–B association model (that is, integer scores version of model (5.20)) does not fit as well as the uniform interaction model. It has $G^2 = 10.8$ based on df $= 6$. However, it yields the simple interpretation of constant local odds ratios $\exp(\hat{\beta}_1^{SB}) = \exp(0.115) = 1.12$ for the age <40 group and $\exp(\hat{\beta}_2^{SB}) = \exp(0.781) = 2.18$ for the 40 to 59 age group. Since $\hat{\beta}_1^{SB} > 0$ and $\hat{\beta}_2^{SB} > 0$, the breathing test results tend to be more abnormal when an individual's smoking status is more current. The association is estimated to be much stronger for the older age group. In fact there is not strong evidence of an association for the younger age group, as $\hat{\beta}_1^{SB} = 0.115$ has an estimated standard error of 0.086.

Assignment of Scores

For the loglinear models studied in this chapter, it is necessary to assign scores to the levels of ordinal variables. Parameter interpretations are simple when the scores are equally spaced. Sometimes it is not obvious how to assign scores, and the researcher may not wish to assume equal spacings. In such cases it is informative to assign scores a variety of "reasonable" ways to check whether substantive conclusions depend on the actual choice.

An alternative approach to scoring ordinal variables in these models will be discussed in Section 8.1. There, the scores in the models are treated as parameters to be estimated from the data rather than numbers to be supplied by the researcher. For further discussion of implications of assigning scores, see Section 12.2.

NOTES

Section 5.1

1. The discussion in the introduction about the inadequacy of the independence model is based on an informal "eyeballing" of Tables 4.7 and 4.11, but it can be more formally quantified. For example, we could assign ordered scores $\{v_j\}$ to the categories of dumping severity and compute a contrast

$$C_i = \sum_j (v_j - \bar{v})\hat{\lambda}_{ij}^{OD}$$

of the association parameter estimates for each operation. For the parameter estimates in Table 4.11 and the scores $v_1 = 1$, $v_2 = 2$, $v_3 = 3$, the contrasts equal $C_1 = -0.517$, $C_2 = -0.007$, $C_3 = 0.072$, $C_4 = 0.450$. The monotone increase in the $\{C_i\}$ reflects the tendency for greater dumping severity to accompany greater operation severity. We could also utilize the ordinal nature of operation severity by assigning ordered scores $\{u_i\}$ to the rows and computing the single contrast $C = \sum\sum(u_i - \bar{u})(v_j - \bar{v})\hat{\lambda}_{ij}^{OD}$, which is positive in this case. Since these contrasts are linear functions of the $\{\hat{\lambda}_{ij}^{OD}\}$, their standard errors can be obtained using the covariance matrix of the $\{\hat{\lambda}_{ij}^{OD}\}$. See Goodman (1971).

2. Yet another statistic for testing independence using model (5.1) can be based on dividing the "score" statistic $\sum\sum u_i v_j(p_{ij} - p_{i+}p_{+j})$ by its standard error, which is

$$\sqrt{[\sum u_i^2 p_{i+} - (\sum u_i p_{i+})^2][\sum v_j^2 p_{+j} - (\sum v_j p_{+j})^2]/n}$$

The score statistic is obtained by differentiating the log likelihood function with respect to β, and evaluating it at $\beta = 0$ and at the independence model ($\beta = 0$) estimates of the other parameters. This test statistic, unlike the others, can be computed without fitting the model. Under H_0 and for local alternatives, it follows from Cox and Hinkley (1974, pp. 322–324) that this test is asymptotically equivalent to the ones based on $G^2(I \mid U)$ and on $\hat{\beta}$.

Section 5.2

3. The asymptotic covariance matrix of parameter estimates can be used to make pairwise comparisons of $\{\tau_i\}$ parameters from the row effects model. For example, one can test H_0: $\tau_a = \tau_b$ by dividing $\hat{\tau}_b - \hat{\tau}_a$ by its asymptotic standard error,

$$\sqrt{\sigma^2(\hat{\tau}_a) + \sigma^2(\hat{\tau}_b) - 2\ \text{cov}\ (\hat{\tau}_a, \hat{\tau}_b)}$$

The ratio has an asymptotic standard normal distribution if H_0 is true. Standard errors of estimates can also be used to construct confidence intervals for association parameters, in the same way as was shown at the end of Section 5.1 for β in the linear-by-linear association model.

COMPLEMENTS

1. Suppose that the sample data for a 3×3 table have the following symmetric pattern around the middle column:

15	10	15
10	10	10
5	20	5

   a. Show that the uniform association model would give $\hat{\beta} = 0$ for these data. (Note that the $\{\hat{m}_{ij}\}$ must have constant local odds ratio but, from the likelihood equations, a correlation of zero for this table.)

   b. Show that for these data the $\{\hat{m}_{ij}\}$ are identical for the independence model and the uniform association model, so that $G^2(I) = G^2(U)$, but $G^2(I \mid U) = 0$. Why does it not make sense to use the conditional test for this pattern of frequencies?

   c. Argue that independence implies $\beta = 0$ in the uniform association model, but $\beta = 0$ does not imply independence unless the model holds.

   d. What would be the results of fitting the row effects model with $\{v_j = j\}$ to these data?

2. Is it necessarily true that all $\log \theta_{ij}$ have the same sign:

   a. If the linear-by-linear association model holds?

   b. If the row effects model holds?

3. For the loglinear row effects model show that, without loss of generality, one can let the parameters satisfy $\sum \lambda_i^X = \sum \lambda_j^Y = \sum \tau_i = 0$.

4. Show that the row effects model is equivalent to the linear-by-linear association model when $r = 2$.

5. Suppose that there is full multinomial sampling over the cells of a two-way cross-classification table. The kernel of the log-likelihood function (see Complement 2.5 and Note 4.1) is $\sum \sum n_{ij} \log m_{ij}$:

a. Suppose that the columns of the table are ordered. Substitute the row effects model for $\log m_{ij}$ in the log likelihood and show that the minimal sufficient statistics for the parameters are $\{n_{i+}\}$, $\{n_{+j}\}$, and $\{\sum_j n_{ij} v_j\}$.

b. Show that the likelihood equations for the ML estimates are $\hat{m}_{i+} = n_{i+}$, $\hat{m}_{+j} = n_{+j}$, and $\sum_j \hat{m}_{ij} v_j = \sum_j n_{ij} v_j$.

c. Let $p_{j(i)} = n_{ij}/n_{i+}$ and $\hat{\pi}_{j(i)} = \hat{m}_{ij}/\hat{m}_{i+}$ denote the estimates of the Y conditional probability $\pi_{j(i)} = \pi_{ij}/\pi_{i+}$ under the saturated model (i.e., for the observed data) and under the row effects model, respectively. Show that the third likelihood equation can be expressed as $\sum_j v_j p_{j(i)} = \sum_j v_j \hat{\pi}_{j(i)}$. In each row, in other words, the mean of the conditional distribution across the columns is the same when the distribution as based on the estimated expected frequencies as it is when based on the observed data.

6. An alternative model for the ordinal-ordinal table, one that has unknown row *and* column effects, is

$$\log m_{ij} = \mu + \lambda_i^X + \lambda_j^Y + \beta(u_i - \bar{u})(v_j - \bar{v}) + \tau_{1i}(v_j - \bar{v}) + \tau_{2j}(u_i - \bar{u})$$

The version of this model having integer scores was proposed by Goodman (1979a, 1981a), who referred to it as "Model I" or as the "$R + C$ model."

a. Give constraints that the model parameters can be assumed to satisfy and that reflect their linear dependencies.

b. Show that $\mathrm{df} = (r - 2)(c - 2)$ for testing goodness of fit of the model.

c. For this model applied with integer scores, show that the local log odds ratio can be expressed in the additive form

$$\log \theta_{ij} = \gamma_i + \delta_j$$

(A model with the same df but having *multiplicative* row and column effects is considered in Section 8.1)

d. Show that the linear-by-linear association model and the row effects model are special cases of this model.

7. For the ordinal-ordinal table, consider the model

$$\log m_{ij} = \mu + \lambda_i^X + \lambda_j^Y + \beta_1(u_i - \bar{u})(v_j - \bar{v}) + \beta_2(u_i - \bar{u})^2(v_j - \bar{v})$$

a. Describe the association pattern represented by this model.

b. For a $3 \times c$ table, how does this model compare to the row effects model?

c. Show that the general association term λ_{ij}^{XY} for the two-way table can be replaced by an expansion involving orthogonal polynomials. (See Haberman 1974b.)

8. Show that the likelihood equations for the following:

a. Simple association model (5.11) are

$$\sum_i \sum_j u_i v_j n_{ij+} = \sum_i \sum_j u_i v_j \hat{m}_{ij+},$$

$$\sum_j \sum_k v_j w_k n_{+jk} = \sum_j \sum_k v_j w_k \hat{m}_{+jk}$$

$$\sum_i \sum_k u_i w_k n_{i+k} = \sum_i \sum_k u_i w_k \hat{m}_{i+k}$$

b. Simple interaction model (5.17) are those in (a) plus

$$\sum_i \sum_j \sum_k u_i v_j w_k n_{ijk} = \sum_i \sum_j \sum_k u_i v_j w_k \hat{m}_{ijk}$$

c. Interaction model (5.19) are

$$n_{ij+} = \hat{m}_{ij+}, \quad n_{+jk} = \hat{m}_{+jk}, \quad n_{i+k} = \hat{m}_{i+k}, \quad \text{all } i, j, k, \quad \text{and}$$

$$\sum_i \sum_j \sum_k u_i v_j w_k n_{ijk} = \sum_i \sum_j \sum_k u_i v_j w_k \hat{m}_{ijk}$$

d. Interpret the likelihood equations for these models.

9. The following table, taken from the 1974 and 1975 General Social Surveys, appears in Haberman (1978, p. 303). The question asked was, Do you believe that women should take care of running their homes and leave running the country up to men? For men respondents, the independence model applied to the 3×2 cross-classification of education and response yields $G^2 = 139.50$, whereas the uniform association model yields $G^2 = 1.93$. For women respondents, $G^2 = 230.94$ for the independence model and $G^2 = 0.79$ for the uniform association model.

a. Comment on the goodness of fit of the independence and uniform association models.

b. For each sex, test whether response is independent of education, given that the uniform association model fits.

c. The standard no three-factor interaction model applied to the three-dimensional cross-classification of R = response, S = sex, and E = education (i.e., the structure (RS, RE, ES)) yields $G^2 =$

8.72 based on df = 2. Thus none of the standard unsaturated loglinear models fits the data. Show that the heterogeneous uniform *R–E* association model (5.20) is an unsaturated three-factor interaction model that fits the data by exploiting the ordinal nature of education.

d. The estimated constant odds ratio is 2.67 when the uniform association model is used for the male sample, and it is 3.94 when used for the female sample. Use these odds ratios and the percent agree values in the table to explain the meaning of the interaction among the three variables.

		Response		
Sex	Education	Agree	Disagree	Percent agree
Male	≤ 8	161	90	64.1
	9–12	212	378	35.9
	≥ 13	92	372	19.8
Female	≤ 8	169	67	71.6
	9–12	325	567	36.4
	≥ 13	61	377	13.9

Source: Haberman (1978). Reprinted by permission.

10. The following table, taken from Duncan and McRae (1979), is a cross-classification of X = race, Y = year, and Z = rated performance of radio and TV networks. The standard loglinear model (XY, XZ, YZ) fits reasonably well ($G^2 = 3.57$, df = 2) but does not utilize the ordinal nature of Z.

a. The homogeneous uniform association model (5.11) with integer scores has $G^2 = 5.58$ based on df = 4. Interpret the pairwise associations for that model using the estimated conditional local log odds ratios $\hat{\beta}^{XY} = 0.561$, $\hat{\beta}^{XZ} = 0.542$, and $\hat{\beta}^{YZ} = -0.307$.

b. The uniform interaction model (5.17) with integer scores has $G^2 =$ 2.22 based on df = 3 and gives a more satisfactory fit than either of the preceding models. The estimated parameters for that model are $\hat{\beta}^{XY} = 0.812$, $\hat{\beta}^{XZ} = 0.671$, $\hat{\beta}^{YZ} = -0.462$, and $\hat{\beta}^{XYZ} = -0.405$. Show that the tendency of blacks to rate performance of the networks higher than do whites is greater in 1959 than in 1971.

Year	Race	Rated Performance of Radio and TV Networks		
		Poor	Fair	Good
1959	White	54	253	325
	Black	4	23	81
1971	White	158	636	600
	Black	24	144	224

Source: Duncan and McRae (1979). Reprinted by permission.

CHAPTER 6

Logit Models

For the loglinear models discussed in Chapters 3 through 5, it is unnecessary to distinguish between response variables and explanatory variables. Expected frequencies are modeled in terms of associations and interactions among *all* the variables. By contrast, the "logit" type of model discussed in the next two chapters describes effects of a set of explanatory variables on a response variable. Like regression models for quantitative response variables, logit models do not describe association and interaction patterns among explanatory variables. Thus in multivariable situations they are simpler to formulate than loglinear models.

In the first section of this chapter we introduce the logit transformation and describe its use in models for dichotomous response variables. When they are applied to cross-classifications having dichotomous response variables, logit models have a simple relationship with previously studied loglinear models. Section 6.2 discusses this correspondence. In Section 6.3 logit models are formulated for cross-classifications having ordinal (rather than dichotomous) response variables.

6.1. LOGISTIC REGRESSION

Let Y denote a response variable that can assume only two values, say 0 and 1. Denote the expected value of Y by

$$E(Y) = P(Y = 1) = \pi$$

and suppose that we want to model the way $\pi = \pi(\mathbf{x})$ depends on the values of explanatory variables $\mathbf{X} = (X_1, \ldots, X_k)$. The standard linear regression model has the form

$$\pi(\mathbf{x}) = \alpha + \beta_1 x_1 + \cdots + \beta_k x_k \tag{6.1}$$

There are several difficulties with using ordinary least squares to fit a model of the form (6.1). Conditions that make least squares estimates optimal are not satisfied here. For instance, the variance of Y is $\pi(\mathbf{x})(1 - \pi(\mathbf{x}))$, which is not constant over the range of values of the explanatory variables. Standard distributional statements for estimators do not apply, since Y is dichotomous rather than normally distributed.

A weighted least squares approach can be used to obtain more efficient estimates of the regression parameters in this model. See, for example, Neter and Wasserman (1974, pp. 326–328) and Draper and Smith (1981, pp. 108–116). The model itself is likely to be inaccurate in certain regions, however, if some X_i are quantitative. This follows because the model predicts the impossible values $\pi < 0$ and $\pi > 1$ for sufficiently large or sufficiently small values of X_i. For a dichotomous response, $E(Y)$ cannot be linearly related to X_i over an unbounded range of X_i values.

Logistic Function

Because of the factors just discussed, it is often more appropriate to use a model that allows a curvilinear relationship between $E(Y)$ and each quantitative X_i. For the case of a single explanatory variable X, the S-shaped curves shown in Figure 6.1 are natural shapes for regression curves if we expect a monotonic relationship.

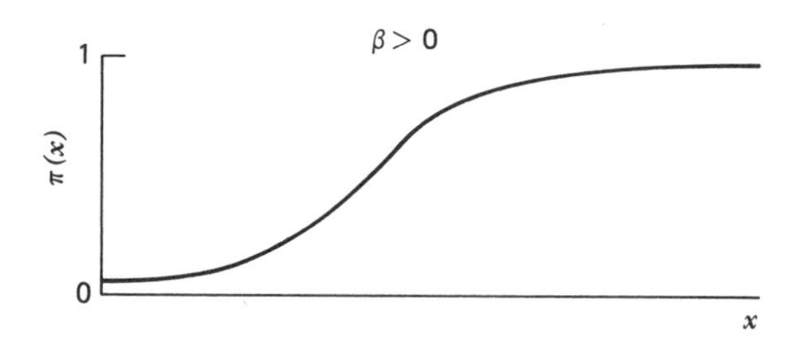

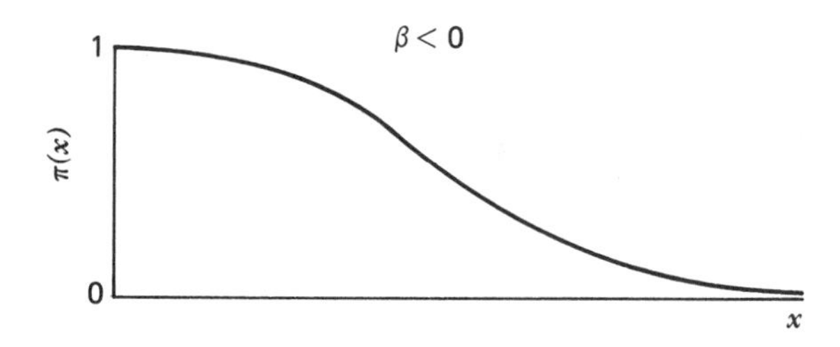

Figure 6.1. Logistic functions.

One function that has this appearance is

$$\pi(x) = \frac{\exp(\alpha + \beta x)}{[1 + \exp(\alpha + \beta x)]} \tag{6.2}$$

called the *logistic function.* This function is monotonic, with $\pi(x) \downarrow 0$ or $\pi(x) \uparrow 1$ as $x \uparrow \infty$ depending on whether $\beta < 0$ or $\beta > 0$. It takes the value $\pi(x) = \frac{1}{2}$ at $x = -\alpha/\beta$, and the curve has a steeper rate of increase around that value as $|\beta|$ increases. For example, $1/\beta$ is the approximate distance between the x values where $\pi(x) = 0.25$ or 0.75 and where $\pi(x) = 0.50$. Greater intuition can be obtained regarding this curve by noting that when $\beta > 0$, it is the distribution function of the logistic random variable having mean $-\alpha/\beta$ and standard deviation $\pi/\sqrt{3}\beta$,

For relationship (6.2), the odds of making response 1 instead of response 0 is

$$\frac{\pi(x)}{(1 - \pi(x))} = \exp(\alpha + \beta x)$$

The odds increases multiplicatively by e^{β} for every unit increase in x. The log odds has the simple linear relationship

$$\log\left[\frac{\pi(x)}{(1 - \pi(x))}\right] = \alpha + \beta x \tag{6.3}$$

The log odds transformation is referred to as the *logit*, and a model for the log odds is called a logit (or logistic) regression model. If relationship (6.2) holds, the logit transformation yields a linear relationship for the logit model. When there are several explanatory variables, the logit regression model generalizes to

$$\log\left[\frac{\pi(\mathbf{x})}{(1 - \pi(\mathbf{x}))}\right] = \alpha + \beta_1 x_1 + \cdots + \beta_k x_k$$

Logit Models for Categorical Data

Now suppose that all variables are categorical. We illustrate the formulation of logit models by considering the three-dimensional case in which one of the variables (say Z) is a dichotomous response variable. As before, let $\{\pi_{ijk}\}$ denote cell probabilities, and let $\{m_{ijk}\}$ denote expected cell frequencies. At levels i of X and j of Y, the conditional probability of making response k for Z is $\pi_{k(ij)} = \pi_{ijk}/\pi_{ij+}$. The logit for Z at that combination of levels of X and Y is

$$\log\left[\frac{\pi_{2(ij)}}{(1-\pi_{2(ij)})}\right] = \log\left(\frac{\pi_{2(ij)}}{\pi_{1(ij)}}\right)$$
$$= \log\left(\frac{\pi_{ij2}}{\pi_{ij1}}\right)$$
$$= \log\left(\frac{m_{ij2}}{m_{ij1}}\right) \tag{6.4}$$

Consider first the case that X and Y are both nominal. A model that has additive effects of X and Y on the logit is

$$\log\left(\frac{m_{ij2}}{m_{ij1}}\right) = \alpha + \tau_i^X + \tau_j^Y \tag{6.5}$$

where $\sum \tau_i^X = \sum \tau_j^Y = 0$. The $\{\tau_i^X\}$ pertain to the partial association between X and Z, and the $\{\tau_j^Y\}$ pertain to the partial association between Y and Z. If model (6.5) holds and if all $\tau_i^X = 0$, then Z is conditionally independent of X, given Y. Model (6.5) assumes an absence of three-factor interaction, since the effects of each explanatory variable are the same at each level of the other one. This model resembles the two-way analysis of variance (ANOVA) model that assumes no interaction. The ANOVA model also has additive effects of qualitative explanatory variables [as on the right-hand side of equation (6.5)], but it is designed for a continuous instead of dichotomous response variable.

Suppose that X is nominal but that Y is ordinal and has monotone scores $\{v_1, \ldots, v_c\}$ assigned to its levels. A simple logit model in which the effect of the ordinal variable on the logit for Z is linear is given by

$$\log\left(\frac{m_{ij2}}{m_{ij1}}\right) = \alpha + \tau_i^X + \beta^Y(v_j - \bar{v}) \tag{6.6}$$

where $\sum \tau_i^X = 0$. This model also assumes an absence of three-factor interaction. We shall discuss interpretations of the parameters in these models in the next section.

In fitting logit models, it is common to treat as fixed (rather than random) the cell counts for the marginal table consisting of combinations of levels of the explanatory variables. For models (6.5) and (6.6), for instance, the $\{n_{ij+}\}$ are regarded as fixed. For all logit models considered in this text, we assume an independent multinomial distribution on the response at each combination of levels of explanatory variables.

Logit models for cross-classifications can be fitted using weighted least squares, as illustrated by Grizzle et al. (1969) and by Theil (1970). Logit models can also be fitted using maximum likelihood, as shown by Haberman (1978, Ch. 5) and by Walker and Duncan (1967) and Cox (1970) for the case where explanatory variables may be continuous. Many computer packages contain programs for fitting them (e.g., SAS-LOGIST, BMDP-LR, SPSS^{X}-LOGLINEAR) or can be adapted to fit them (e.g., GLIM).

The degrees of freedom for chi-squared goodness-of-fit tests of logit models are easily calculated. They equal the number of logits that can be formed in the table minus the number of linearly independent parameters in the model. For the three-way $r \times c \times 2$ table with dichotomous response Z, there are $r \times c$ logits, one at each combination of levels of X and Y. For model (6.5) therefore

$$\text{df} = rc - [1 + (r - 1) + (c - 1)] = (r - 1)(c - 1)$$

whereas for model (6.6)

$$\text{df} = rc - [1 + (r - 1) + 1] = r(c - 1) - 1$$

6.2. LOGIT MODELS AND LOGLINEAR MODELS

We have seen that loglinear models describe cell expected frequencies in terms of association and interaction patterns among all variables. Logit models describe the effects of explanatory variables on a dichotomous response variable, and they disregard structural relationships among the explanatory variables. Although these two model types seem quite distinct, it is possible to obtain the logit model from a corresponding loglinear model. That loglinear model contains the same structure as the logit model for the associations between the response variable and the explanatory variables, and it contains the most general interaction term for describing relationships among the explanatory variables.

Correspondence between Models

To illustrate, we construct the loglinear model corresponding to logit model (6.5). In model (6.5) there is no three-factor interaction but the response Z is associated with the nominal variables X and Y. Hence we include the association terms λ_{ik}^{XZ} and λ_{jk}^{YZ} in the loglinear model. We also include the general association term λ_{ij}^{XY} for the relationship between the two explanatory variables. This forces the fitted X–Y marginal frequencies to equal the observed ones, and is desirable since we treat these values as fixed in the independent multinomial sampling scheme for the logit model. The resulting loglinear model is the one symbolized by (XY, XZ, YZ), namely

$$\log m_{ijk} = \mu + \lambda_i^X + \lambda_j^Y + \lambda_k^Z + \lambda_{ij}^{XY} + \lambda_{ik}^{XZ} + \lambda_{jk}^{YZ} \tag{6.7}$$

To show that logit model (6.5) is implied by loglinear model (6.7), we consider the logit

$$\log\left(\frac{m_{ij2}}{m_{ij1}}\right) = \log(m_{ij2}) - \log(m_{ij1})$$

From (6.7) this difference is

$$\begin{aligned} &[\mu + \lambda_i^X + \lambda_j^Y + \lambda_2^Z + \lambda_{ij}^{XY} + \lambda_{i2}^{XZ} + \lambda_{j2}^{YZ}] \\ &- [\mu + \lambda_i^X + \lambda_j^Y + \lambda_1^Z + \lambda_{ij}^{XY} + \lambda_{i1}^{XZ} + \lambda_{j1}^{YZ}] \\ &= (\lambda_2^Z - \lambda_1^Z) + (\lambda_{i2}^{XZ} - \lambda_{i1}^{XZ}) + (\lambda_{j2}^{YZ} - \lambda_{j1}^{YZ}) \end{aligned}$$

Now, $\lambda_1^Z = -\lambda_2^Z$, $\lambda_{i1}^{XZ} = -\lambda_{i2}^{XZ}$, and $\lambda_{j1}^{YZ} = -\lambda_{j2}^{YZ}$, since Z is dichotomous and since $\sum_k \lambda_k^Z = \sum_k \lambda_{ik}^{XZ} = \sum_k \lambda_{jk}^{YZ} = 0$. Thus the logit simplifies to

$$\log\left(\frac{m_{ij2}}{m_{ij1}}\right) = 2\lambda_2^Z + 2\lambda_{i2}^{XZ} + 2\lambda_{j2}^{YZ} \tag{6.8}$$

This is precisely the form of the logit model (6.5) if we identify $2\lambda_{i2}^{XZ}$ as the ith effect of X on the logit of Z (i.e., τ_i^X), $2\lambda_{j2}^{YZ}$ as the jth effect of Y on the logit of Z (i.e., τ_j^Y), and $2\lambda_2^Z = \alpha$.

The λ_{ij}^{XY} terms regarding association among explanatory variables cancel in the difference in logs represented by the logit. Information about this association is not included in the logit model.

Parameter Interpretations

From (6.8), it follows that the parameters in logit model (6.5) satisfy

$$\sum_i \tau_i^X = 2\sum_i \lambda_{i2}^{XZ} = 0 \quad \text{and} \quad \sum_j \tau_j^Y = 2\sum_j \lambda_{j2}^{YZ} = 0$$

Therefore

$$\sum_i \sum_j \log\left(\frac{m_{ij2}}{m_{ij1}}\right) = 2rc\lambda_2^Z = rc\alpha$$

and $\alpha = (1/rc) \sum_i \sum_j \log(m_{ij2}/m_{ij1})$ can be interpreted as the mean of the $r \times c$ logits. We can regard τ_i^X as a deviation from the mean due to location of the cell in row i, and we can regard τ_j^Y as a deviation from the mean due to location in column j.

When an explanatory variable is also dichotomous, its association parameter in the logit model has a simple odds ratio interpretation. For example, suppose that Y is dichotomous. Consider the conditional log odds ratios for Y and the response Z, controlling for X. For $i = 1, \ldots, r$, log $[(m_{i11}m_{i22})/(m_{i12}m_{i21})]$ is the difference between the logits, $\log(m_{i22}/m_{i21}) - \log(m_{i12}/m_{i11})$. For model (6.5) this difference equals

$$(\alpha + \tau_i^X + \tau_2^Y) - (\alpha + \tau_i^X + \tau_1^Y) = \tau_2^Y - \tau_1^Y = 2\tau_2^Y$$

Therefore $2\tau_2^Y$ can be regarded as the log odds ratio between Y and Z at each level of X. This measure also equals $4\lambda_{22}^{YZ} = 4\lambda_{11}^{YZ}$ from the corresponding loglinear model, and [as noted in (4.2)] the loglinear parameter λ_{11}^{YZ} can be interpreted as one-fourth the conditional Y–Z log odds ratio.

If an explanatory variable is ordinal, and if it has the linear effect indicated in model (6.6), then its β parameter has a slope interpretation. In other words, β represents the change in the logit, and e^{β} represents the multiplicative change in the odds, per unit change in the ordinal variable.

Since logit models for cross-classification tables are directly related to loglinear models, another way to fit them is to fit the corresponding loglinear model and use the relationship between the parameters of the two models. Hence existing computer packages (see Appendix D) for fitting loglinear models can also be used to fit logit models. The residual degrees of freedom for the two model types match properly when the loglinear model contains the general form for interactions among explanatory variables, since no assumption is made about the form of their relationship in the logit model. For example, loglinear model (6.7) has residual $\text{df} = (r-1)(c-1)(l-1)$, which simplifies to $(r-1)(c-1)$ when $l = 2$. This is the value previously obtained for logit model (6.5). See also Complement 6.2.

Death Penalty Data

In the loglinear analyses of Radelet's data on racial characteristics and the death penalty (see Table 3.1 and Sections 4.1–4.3), we did not identify a response variable. However, one would normally regard P = death penalty verdict as the response and D = defendant's race and V = victim's race as explanatory variables. The logit model that includes both explanatory effects but no three-factor interaction terms is

$$\log\left(\frac{m_{ij2}}{m_{ij1}}\right) = \alpha + \tau_i^D + \tau_j^V$$

There are four logits, one for each combination (i, j) of D and V categories. The model contains three linearly independent parameters, since $\tau_1^D = -\tau_2^D$ and $\tau_1^V = -\tau_2^V$. Its residual degrees of freedom equals $4 - 3 = 1$.

For the corresponding loglinear model (DP, VP, DV), the ML parameter estimates are $\hat{\lambda}_2^P = 1.20$, $\hat{\lambda}_{22}^{DP} = -0.11$, and $\hat{\lambda}_{22}^{VP} = 0.33$. For the logit model therefore

$$\hat{\alpha} = 2\hat{\lambda}_2^P = 2.40$$

$$\hat{\tau}_2^D = 2\hat{\lambda}_{22}^{DP} = -0.22 = -\hat{\tau}_1^D$$

$$\hat{\tau}_2^V = 2\hat{\lambda}_{22}^{VP} = 0.66 = -\hat{\tau}_1^V$$

The asymptotic standard errors of $\hat{\tau}_2^D$ and $\hat{\tau}_2^V$ are 0.20 and 0.26, respectively, twice those of $\hat{\lambda}_{22}^{DP}$ and $\hat{\lambda}_{22}^{VP}$.

The estimated average logit $\hat{\alpha}$ is positive, indicating an overall tendency for there to be more "no" outcomes (level 2 of P) than "yes" outcomes on the death penalty verdict. Classification of defendant's race as "black" (level 2 of D) has a negative effect on the logit for death penalty verdict. Thus it is estimated that black defendants have a smaller chance than white defendants of receiving the "no" outcome on death penalty verdict, controlling for victim's race. The effect on the logit is positive when victim's race is classified black. The estimated D–P and V–P partial odds ratios are $\exp(2\hat{\tau}_2^D) = 0.66$ and $\exp(2\hat{\tau}_2^V) = 3.45$.

The logit model fits well, since the value of $G^2 = 0.70$ based on df $= 1$ is necessarily the same as the value obtained for testing the goodness of fit of the loglinear model (DP, VP, DV). The simpler loglinear model (VP, DV) that was also observed to fit well corresponds to the simpler logit model

$$\log(m_{ij2}/m_{ij1}) = \alpha + \tau_j^V.$$

Heart Disease Example

Table 6.1, taken from Ku and Kullback (1974), is based on data originally reported by Cornfield (1962). A sample of male residents of Framingham, Massachusetts, aged 40 to 59, was classified on blood pressure (P) and serum cholesterol (C) levels. During a six-year follow-up period, they were also classified according to whether they developed coronary heart disease (H). Here it is natural to treat the dichotomous variable H as a response variable and the ordinal variables C and P as explanatory variables.

The logit model

$$\log\left(\frac{m_{ij2}}{m_{ij1}}\right) = \alpha + \beta^C(u_i - \bar{u}) + \beta^P(v_j - \bar{v})$$

assumes an absence of three-factor interaction. It has $4 \times 4 = 16$ logits and 3 parameters, so its residual degrees of freedom equals 13. The correspond-

Table 6.1. Cross-Classification of Framingham Men by Blood Pressure, Serum Cholesterol, and Presence or Absence of Heart Disease

Coronary Heart Disease	Serum Cholesterol (mg/100 cc)	Systolic Blood Pressure (mm Hg)			
		<127	127–146	147–166	167+
Present	<200	2	3	3	4
	200–219	3	2	0	3
	220–259	8	11	6	6
	≥260	7	12	11	11
	<200	117	121	47	22
Absent	200–219	85	98	43	20
	220–259	119	209	68	43
	≥260	67	99	46	33

Source: Ku and Kullback (1974). Reprinted by permission.

ing loglinear model has linear-by-linear C–H and P–H associations, and a general C–P association term. It is

$$\log m_{ijk} = \mu + \lambda_i^C + \lambda_j^P + \lambda_k^H + \lambda_{ij}^{CP}$$
$$+ \beta^{CH}(u_i - \bar{u})(w_k - \bar{w}) + \beta^{PH}(v_j - \bar{v})(w_k - \bar{w})$$

If the scores for the levels of H satisfy $w_2 - w_1 = 1$, then β^C and β^H in the logit model are identical to β^{CH} and β^{PH} in the loglinear model, respectively.

For integer scoring of these models, β^C represents a uniform local log odds ratio for the partial C–H association, and β^P represents a uniform local log odds ratio for the partial P–H association. For the ML fitting of either model, $\hat{\beta}^C = 0.530$ and $\hat{\beta}^P = 0.440$. The odds of developing coronary heart disease are estimated to be $\exp(0.530) = 1.70$ times higher if the cholesterol level is $i + 1$ instead of i. Each increment in blood pressure level increases the odds by a factor of $\exp(0.440) = 1.55$. The asymptotic standard errors of $\hat{\beta}^C$ and $\hat{\beta}^P$ are 0.117 and 0.109, respectively, so the contribution of each term is highly significant. The model fits quite well, with $G^2 = 14.8$ based on df $= 13$.

6.3. LOGITS FOR AN ORDINAL RESPONSE

The logit models of Sections 6.1 and 6.2 were formulated for dichotomous response variables. If there are c response categories with $c \geq 2$, there are many ways of forming logits. For example, for a polytomous variable having response probabilities $(\pi_1, \ldots, \pi_c)$ at a certain combination of levels of explanatory variables, we could form the conditional logit

$$\log\left[\frac{(\pi_j/(\pi_j+\pi_k))}{(\pi_k/(\pi_j+\pi_k))}\right] = \log\left(\frac{\pi_j}{\pi_k}\right).$$

This is the log odds of classification in category j instead of category k, given that an observation falls in one of those two basic categories. All logits of this type can be derived from a basic set of size $c-1$. For example, if

$$L_j = \log(\pi_j/\pi_c), \qquad j = 1, \ldots, c-1 \tag{6.9}$$

then $\log(\pi_j/\pi_k) = L_j - L_k$ for $1 \le j < k \le c-1$. Models for the $\{L_j\}$ can also be fitted using corresponding loglinear models (see Fienberg 1980, Sec. 6.6). Anderson's (1984) "stereotype" model uses the $\{L_j\}$ in models for ordinal response variables, by constraining corresponding regression parameters to have a monotonic pattern.

Forming Logits for Ordinal Variables

When the response variable is ordinal it makes sense to form logits in a way that takes the category order into account. In so doing, it is not necessary to use only two categories at a time in forming the logits, as was done in (6.9). The logits can be formed by grouping categories that are contiguous on the ordinal scale, for example. Among the logit types that utilize category ordering are the "cumulative" or "accumulated" logits

$$L_j = \log\left[\frac{(\pi_{j+1}+\cdots+\pi_c)}{(\pi_1+\cdots+\pi_j)}\right], \qquad j = 1, \ldots, c-1 \tag{6.10}$$

the "continuation-ratio" logits

$$L_j = \log\left[\frac{\pi_{j+1}}{(\pi_1+\cdots+\pi_j)}\right], \qquad j = 1, \ldots, c-1$$

and the "adjacent-categories" logits

$$L_j = \log\left[\frac{\pi_{j+1}}{\pi_j}\right], \qquad j = 1, \ldots, c-1$$

When $c = 2$, all three of these logit types simplify to the standard logit $\log(\pi_2/\pi_1)$.

The adjacent-categories logits are a basic set equivalent to (6.9). Models using these logits were presented by Goodman (1983). His models are equivalent to loglinear models discussed in the previous chapter, but they have different emphasis in the sense that one variable is treated as a response. See Complement 7.8 for the formulation of logit models for adjacent categories and for the equivalences with ordinal loglinear models.

Continuation-ratio logits have the appealing feature that the results of fitting models for separate logits are independent (see Complement 6.3). Hence the $c - 1$ G^2 statistics and their df values can be summed to obtain an overall goodness-of-fit statistic that pertains to the simultaneous fitting of $c - 1$ models, one for each logit. However, the logits that are formed (and hence the results of the analysis) differ for the two possible orders in which the categories of the ordinal variable can be listed. The alternative logits

$$L'_j = \log\left[\frac{\pi_j}{(\pi_{j+1} + \cdots + \pi_c)}\right], \qquad j = 1, \ldots, c - 1,$$

can be interpreted as log odds for "hazard" probabilities, since L'_j uses the ratio of the conditional probabilities $\pi_j/(\pi_j + \cdots + \pi_c)$ and $(\pi_{j+1} + \cdots + \pi_c)/(\pi_j + \cdots + \pi_c)$. In fact, Thompson (1977) proposed a model having these logits for the analysis of discrete survival-time data. He showed that when the lengths of the time intervals approach zero, his model converges to Cox's (1972) proportional hazards model for survival data. Fienberg and Mason (1979) used continuation-ratio logits in age-period-cohort models for the analysis of discrete archival data. For other examples of their use, see Fienberg (1980, pp. 114–116) and McCullagh and Nelder (1983, p. 115).

Cumulative logits have been used by several authors. The article by McCullagh (1980) is a good one for motivating their use. In the remainder of this chapter and in Chapter 7, we shall formulate logit models for ordinal response variables using this type of logit. The cumulative logits use all c categories for each logit. As we have defined them, they are the negatives of logits of distribution function values; that is,

$$L_j = -\log\left[\frac{F_j}{(1 - F_j)}\right], \quad \text{where} \quad F_j = \sum_{i \leq j} \pi_i$$

Therefore the $\{L_j\}$ necessarily satisfy $L_1 \geq L_2 \geq \cdots \geq L_{c-1}$.

To illustrate the use of a logit model for an ordinal response variable, we now form a cumulative logit model for a two-way table. Suppose that the column variable is an ordinal response, and suppose that a fixed point j is selected for the cut point for the cumulative logit. The jth cumulative logit in row i is

$$L_{j(i)} = \log\left[\frac{(m_{i,j+1} + \cdots + m_{ic})}{(m_{i1} + \cdots + m_{ij})}\right], \qquad i = 1, \ldots, r$$

If the row variable is also ordinal, scores $\{u_i\}$ may be assigned to its levels.

A simple linear model for the jth cumulative logit values $\{L_{j(1)}, L_{j(2)}, \ldots, L_{j(r)}\}$ is then

$$L_{j(i)} = \alpha_j + \beta_j(u_i - \bar{u}), \qquad i = 1, \ldots, r \tag{6.11}$$

There are r logits and two parameters (for fixed j), so that df $= r - 2$.

Model (6.11) is simply the basic logit model for a dichotomous response variable and quantitative explanatory variable. Here the model has been applied to an $r \times 2$ table in which the original response classification is collapsed into two categories. The first category is obtained by combining the first j categories of the response variable, and the second is obtained by combining the last $c - j$ categories. If model (6.11) holds and $\beta_j = 0$, then the two variables are independent when the response is collapsed in this manner.

For model (6.11) the difference between the logits in adjacent rows is $L_{j(i+1)} - L_{j(i)} = \beta_j(u_{i+1} - u_i)$, which simplifies to β_j when integer scores are used. Now $L_{j(i+1)} - L_{j(i)}$ is the log odds ratio obtained for the 2×2 table consisting of rows i and $i + 1$ and the dichotomous response. For integer row scores this model implies that these log odds ratios are the same (i.e., equal to β_j) for all $r - 1$ pairs of adjacent rows. Hence model (6.11) for integer scores and fixed j is identical to the loglinear uniform association model [(5.1) with integer scores] applied to the collapsed $r \times 2$ table. We can fit model (6.11) for fixed j by fitting that loglinear model, as described in Section 5.1.

Dumping Severity Example

We now reconsider the cross-classification of dumping severity and operation (Table 5.1) that was analyzed using an ordinal loglinear model in Section 5.1. We treat dumping severity as the response variable and again assign integer scores to the operations. Table 6.2 gives the results of fitting the two linear logit models of the form (6.11), one for each cumulative logit. This model is equivalent to the standard logit model applied to the two possible collapsings of dumping severity.

The logit model fits well for both cumulative logits. Since $\hat{\beta}_j > 0$ in each case, the estimated logit increases as operation severity increases. Hence the predicted effect of greater stomach removal is relatively more observations at the high end of the dumping severity scale.

The case $j = 1$ refers to the table in which the "slight" and "moderate" categories of dumping severity are pooled into a single category. The G^2 goodness-of-fit statistic in that case is $G^2(I) = 7.83$ for the independence model and $G^2 = 1.49$ for the logit model. There is strong evidence of an association ($\beta_1 \neq 0$), since the difference in G^2 values is 6.34 with df $= 1$.

Table 6.2. Results of Fitting Cumulative Logit Models to Dumping Severity Data (Table 5.1)

	Collapse Response after Category j	
	$j = 1$	$j = 2$
$\hat{\alpha}_j$	-0.89	-2.60
$\hat{\beta}_j$	0.229 (se = 0.091)	0.211 (se = 0.142)
G^2 for logit model	1.49 (df = 2)	0.94 (df = 2)
$G^2(I)$, model with $\beta_j = 0$	7.83 (df = 3)	3.15 (df = 3)
G^2 for testing $\beta_j = 0$	6.34 (df = 1, $P = 0.012$)	2.21 (df = 1, $P = 0.137$)

Specifically, $\hat{\beta}_1 = 0.229$ with standard error (se) 0.091 indicates that operation severity is positively associated with dumping severity when the latter is measured as "some" or "none." The value $\exp(\hat{\beta}_1) = 1.26$ is the estimated common odds ratio for adjacent rows. In other words, the odds that there is some dumping (instead of none) is estimated to be 1.26 times higher for operation $i + 1$ than for operation i, $i = 1, 2, 3$. There is a similar estimated odds ratio for the other collapsing ($\exp(\hat{\beta}_2) = 1.23$), but the standard error of $\hat{\beta}_2$ is larger, and there is not such strong evidence that $\beta_2 > 0$.

It is necessary that $\hat{\alpha}_1 \geq \hat{\alpha}_2$ since the second logit cannot be larger than the first logit. However, the parameters β_1 and β_2 that represent the effects of operation on the logits for dumping severity have very similar estimates for these data. This suggests the replacement of the separate models

$$L_{1(i)} = \alpha_1 + \beta_1(u_i - \bar{u}) \quad \text{and} \quad L_{2(i)} = \alpha_2 + \beta_2(u_i - \bar{u})$$

by a single model having a homogeneous effect parameter, say β. This simpler model is

$$L_{j(i)} = \alpha_j + \beta(u_i - \bar{u}), \qquad i = 1, \ldots, r \quad \text{and} \quad j = 1, \ldots, c - 1 \quad (6.12)$$

An amalgamated logit model of this type for the full $r \times c$ table does not correspond to any loglinear model. The next chapter considers homogeneous-effects logit models for ordinal response variables.

NOTES

Section 6.2

1. Several loglinear models can reduce to the same logit model. For instance, the loglinear model (XZ, YZ) also reduces to logit model (6.5), since the X–Y association is the one that drops out when the logit is

formed. However, the G^2 and df values are the same only if a loglinear model contains the fullest interaction among the variables that are explanatory in the logit model.

Section 6.3

2. McCullagh (1978) calls a method *palindromic invariant* if the results are invariant under a complete reversal of order of categories but not under general permutations of categories. Most methods discussed in this text have this property for categories of ordinal variables. For instance, for the linear-by-linear association model (5.1) a reversal of the categories of one of the variables changes only the sign of $\hat{\beta}$ and does not alter G^2 or substantive conclusions. Models based on continuation-ratio logits are not palindromic invariant.

3. The article by McCullagh (1980) and the subsequent discussion give an interesting exposition of many basic issues involved in analyzing ordinal data. His article and the one by Goodman (1979a) have been highly influential in advancing the use of logit and loglinear models, respectively.

COMPLEMENTS

1. Consider the logit model for a two-dimensional table,

$$\log (m_{i2}/m_{i1}) = \alpha + \tau_i^X$$

where $\sum \tau_i^X = 0$.

a. Show that this model is saturated.

b. Show that the case $\tau_1^X = \tau_2^X = \cdots = \tau_r^X = 0$ is the independence model.

2. Consider model (6.6) for explanatory variables X nominal and Y ordinal.

a. Give the interpretation of β^Y for the case when the model is applied with integer scores.

b. Show that the loglinear model corresponding to logit model (6.6) is

$$\log m_{ijk} = \mu + \lambda_i^X + \lambda_j^Y + \lambda_k^Z + \lambda_{ij}^{XY} + \tau_i^X(w_k - \bar{w}) + \beta^Y(v_j - \bar{v})(w_k - \bar{w})$$

$1 \leq i \leq r$, $1 \leq j \leq c$, $1 \leq k \leq 2$, with scores w_1 and w_2 for Z such that $w_2 - w_1 = 1$.

3. Consider the multinomial probability function for the numbers of observations $(n_1, \ldots, n_c)$ occurring in c categories, when the category probabilities are $\{\pi_i\}$. Let

$$\rho_1 = \pi_1, \qquad \rho_2 = \frac{\pi_2}{(1 - \pi_1)}, \ldots, \qquad \rho_{c-1} = \frac{\pi_{c-1}}{(1 - \pi_1 - \cdots - \pi_{c-2})}$$

   a. Show that the multinomial probability function can be factored as

$$b(n, n_1; \rho_1)b(n - n_1, n_2; \rho_2) \cdots b(n - n_1 - \cdots - n_{c-2}, n_{c-1}; \rho_{c-1})$$

   where $b(n, x; \rho)$ represents the binomial probability of x successes in n trials when the success probability is ρ on each trial.

   b. Note that

$$\log\left[\frac{\rho_1}{(1 - \rho_1)}\right], \qquad \log\left[\frac{\rho_2}{(1 - \rho_2)}\right], \ldots, \qquad \log\left[\frac{\rho_{c-1}}{(1 - \rho_{c-1})}\right]$$

   are logits of the continuation-ratio type. Deduce from the factorization of the multinomial likelihood the result that logit models using continuation ratios can be fitted independently.

4. Refer to the heart disease data in Table 6.1. Using a computer package, fit the homogeneous uniform association model (5.11), obtaining $G^2 = 22.8$ based on df $= 21$. Describe how the estimate $\hat{\beta}^{CP} = 0.10$ for that model provides information not given by the logit model considered in Section 6.2.

5. Refer to Complement 5.9. Let m_{ijk} denote the expected frequency at level i of education (E), level j of sex (S), and level k of response. Consider the model

$$\log(m_{ij2}/m_{ij1}) = \alpha + \beta^E(u_i - \bar{u}) + \tau_j^S + \zeta_j^{SE}(u_i - \bar{u})$$

   where $\sum \tau_j^S = \sum \zeta_j^{SE} = 0$.

   a. Fit this model, using integer scores for education, and explain how to interpret the parameter estimates.

   b. Give the corresponding loglinear model from which this logit model can be obtained.

6. Let $F(x) = \exp(x)/[1 + \exp(x)]$, $-\infty < x < \infty$, denote the distribution function of a logistic random variable. The logistic regression model (6.3) satisfies $\pi(x) = F(\alpha + \beta x)$, or

$$F^{-1}[\pi(x)] = \alpha + \beta x$$

   where $F^{-1}[\pi(x)] = \log[\pi(x)/(1 - \pi(x))]$.

a. Show that the logistic density function is symmetric and unimodal.

b. In fact the logistic density is similar in appearance to the normal density, having somewhat thicker tails. Let Φ denote the standard normal distribution function. Argue that another reasonable model for binary response variables is

$$\Phi^{-1}[\pi(x)] = \alpha + \beta x$$

This model is referred to as a *probit* model. See Finney (1971) for an iterative ML procedure for fitting this model.

CHAPTER 7

Logit Models for an Ordinal Response

In Section 6.3 we described several ways of forming logits for ordinal response variables. For each type of logit applied to a c category response, $c - 1$ nonredundant logits can be formed. The analyses in this chapter incorporate these $c - 1$ logits into a single model. This approach gives more parsimonious and simpler to interpret models than the fitting of $c - 1$ separate models, one for each logit. We shall first formulate this simpler model for two-way tables in which the explanatory variable is ordinal (Section 7.1) or nominal (Section 7.2). In Section 7.3 we consider it for the general case of a multidimensional table having an ordinal response variable. The models are formulated for the cumulative logits, but they also make sense with the other types of logits for ordinal variables.

7.1. LOGIT MODEL FOR ORDINAL-ORDINAL TABLES

Suppose that both variables X and Y in a two-way table are ordinal, and suppose that the column variable Y is a response variable. This setting is analogous to the one for the bivariate linear regression model for continuous response variables. Let $L_{j(i)}$ denote the jth cumulative logit within row i; that is,

$$L_{j(i)} = \log\left[\frac{(m_{i,\,j+1} + \cdots + m_{ic})}{(m_{i1} + \cdots + m_{ij})}\right] \tag{7.1}$$

Ordered scores $\{u_i\}$ are assigned to the rows. It is unnecessary to assign scores to the levels of the response variable, since the cumulative logits $\{L_{1(i)}, L_{2(i)}, \ldots, L_{c-1,\,(i)}\}$ within each row serve as responses. In this respect logit models differ from the loglinear models of Chapter 5, for which scores were assigned to the levels of *all* ordinal variables.

Uniform Association Model

For fixed j, in Section 6.3 we considered the model

$$L_{j(i)} = \alpha_j + \beta_j(u_i - \bar{u}), \qquad i = 1, \ldots, r \tag{7.2}$$

This is the standard logit regression model for a dichotomous response, one response consisting of the first j categories of Y and the other consisting of the last $c - j$ categories of Y. When model (7.2) was applied for $j = 1$ and $j = 2$ to the dumping severity data in Section 6.3, the estimates obtained for β_1 and β_2 were very close. This suggests that we assume $\beta_1 = \beta_2$ when we fit model (7.2) for $j = 1$ and $j = 2$ to those data. More generally, one can combine the $c - 1$ versions of model (7.2), corresponding to the $c - 1$ possible cut points of the response for forming the logit, into a single model in which the same slope parameter β is used for each logit.

Under the assumption that $\beta_1 = \cdots = \beta_{c-1}$, model (7.2) simplifies to

$$L_{j(i)} = \alpha_j + \beta(u_i - \bar{u}), \qquad 1 \leq i \leq r, \qquad 1 \leq j \leq c - 1 \tag{7.3}$$

This is a model for $c - 1$ logits in each of r rows, a total of $r(c - 1)$ logits. The model has a single association parameter (β) and $c - 1$ parameters (the $\{\alpha_j\}$) pertaining to the various cut points for forming the logits. The residual df for fitting the model is therefore df $= r(c - 1) - c = rc - r - c$. This is the same as the residual df for the ordinal loglinear model for this setting, the linear-by-linear association model (5.1). However, the loglinear and cumulative logit models are not equivalent unless $c = 2$.

Interpretation of the parameters in model (7.3) is quite simple. Since $\sum_i L_{j(i)} = r\alpha_j$, α_j is the average of the r logits that are formed when the cut point follows category j. For each i, $L_{1(i)} \geq L_{2(i)} \geq \cdots \geq L_{c-1,(i)}$ so that the $\{\alpha_j\}$ are monotone decreasing. The $\{-\alpha_j\}$ are often interpreted as category boundary scores for the response variable (see Note 8.4). Usually the main parameter of interest is the association parameter, β. Each of the $c - 1$ logits is linearly related to the explanatory variable, with slope β assumed the same for all logits. If the model holds and $\beta = 0$, then (for all j) the jth logit is the same in each row, which implies that X and Y are independent. If $\beta > 0$ the logit increases as X increases, which implies that the conditional Y distributions are stochastically higher at higher values of X.

The parameter β can also be interpreted through log odds ratios. The difference in logits $L_{j(b)} - L_{j(a)}$ is the log odds ratio for the 2×2 table obtained using rows a and b and the dichotomous response having cut

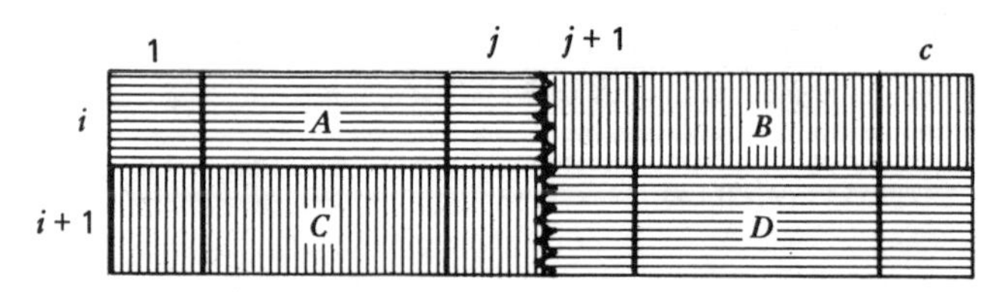

Figure 7.1. Odds ratios AD/BC that are constant for all pairs of adjacent rows and all $c - 1$ cut points for column variable, for logit uniform association model.

point following category j. For model (7.3),

$$L_{j(b)} - L_{j(a)} = \beta(u_b - u_a)$$

Hence the log odds ratio is proportional to the distance between the rows, and (for fixed a and b) it is the same for all cut points.

The difference in logits $L_{j(i+1)} - L_{j(i)}$ for adjacent rows is the log of the local-global odds ratio θ'_{ij} introduced in (2.16). If integer row scores $\{u_i = i\}$ are used in model (7.3), this logit difference equals

$$L_{j(i+1)} - L_{j(i)} = \beta, \qquad 1 \leq i \leq r - 1, \qquad 1 \leq j \leq c - 1 \tag{7.4}$$

In other words, exp (β) then represents the constant value of the odds ratios $\{\theta'_{ij}\}$ for the $(r - 1)(c - 1)$ 2×2 tables obtained by taking all pairs of adjacent rows and all dichotomous collapsings of the response. We shall refer to model (7.3) applied with integer scores as the *logit uniform association model.* Figure 7.1 illustrates the constant odds ratio that is implied by this model.

Estimation of Model

Among the authors who have suggested model (7.3) are Williams and Grizzle (1972) and McCullagh (1980). Williams and Grizzle (1972) fitted the model using weighted least squares (WLS). The computer program GENCAT (Landis et al. 1976) can be used to fit by WLS the models of this chapter. Appendix A.2 includes a discussion of the application of WLS to models of this type. McCullagh suggested using the Newton-Raphson iterative procedure to obtain a maximum likelihood (ML) solution for fitting model (7.3). The general approach is described in Appendix B.3. The computer program MULTIQUAL (Bock and Yates 1973) can be used to fit by ML the models of this chapter.

After estimating the parameters in model (7.3), one can obtain estimated logits and invert them to obtain estimated cell expected frequencies. The conditional distribution function $\{F_{j(i)} = (m_{i1} + \cdots + m_{ij})/(m_{i1} + \cdots + m_{ic}), j = 1, \ldots, c\}$ of the response variable in row i is related to the cumulative logit $L_{j(i)}$ by

$$L_{j(i)} = \log\left[\frac{(1 - F_{j(i)})}{F_{j(i)}}\right], \quad \text{so that}$$

$$F_{j(i)} = [\exp(L_{j(i)}) + 1]^{-1}$$

The estimated logit $\hat{L}_{j(i)} = \hat{\alpha}_j + \hat{\beta}(u_i - \bar{u})$ yields the estimated cumulative probability $\hat{F}_{j(i)} = [\exp(\hat{L}_{j(i)}) + 1]^{-1}$. The jth estimated conditional probability in row i is then

$$\hat{F}_{j(i)} - \hat{F}_{j-1,(i)}, \qquad j = 1, \ldots, c, \quad \text{where} \quad \hat{F}_{0(i)} = 0$$

This estimated probability is multiplied by n_{i+} to get the estimated expected frequency $\hat{m}_{ij}$.

Dumping Severity Example

We now apply this simpler logit model to the data on dumping severity and operation. We use the uniform association version of the model and, as in Section 6.3, treat dumping severity as the response. With the ML approach the estimates of the average logits are $\hat{\alpha}_1 = -0.320$ and $\hat{\alpha}_2 = -2.074$. The estimate of the linear effect of operation on the logit of dumping severity is $\hat{\beta} = 0.225$, with a standard error of .088. Hence the odds that dumping severity is above a certain point rather than below it are estimated to be $\exp(0.225) = 1.25$ times higher for operation $i + 1$ than for operation i, $i = 1, 2, 3$. These results summarize in a simplified manner those given in Table 6.1 for the separate logit models.

The expected frequency estimates for the independence model and for the logit uniform association model are given in parentheses in Table 7.1. To illustrate, for the uniform association model,

$$\hat{L}_{1(1)} = \hat{\alpha}_1 + \hat{\beta}(u_1 - \bar{u}) = -0.320 + 0.225(1 - 2.5) = -0.658$$

which corresponds to $\hat{F}_{1(1)} = [\exp(-0.658) + 1]^{-1} = 0.659$, so that $\hat{m}_{11} = n_{1+}\hat{F}_{1(1)} = 96(0.659) = 63.2$. The fit of the logit model is quite good, with $G^2 = 4.27$ based on df $= 5$.

Conditional Test of Independence

Given that the logit uniform association model holds, one can test independence by testing $H_0: \beta = 0$. A natural test statistic is the difference between the likelihood-ratio G^2 values for the independence model (I) and the logit uniform association model (U); that is

$$G^2(I \mid U) = G^2(I) - G^2(U) \tag{7.5}$$

Table 7.1. Observed Frequencies and Estimated Expected Frequencies for Independence Model and Logit Uniform Association Model, for Dumping Severity Data

	Dumping Severity			
Operation	None	Slight	Moderate	Total
A	61	28	7	96
	(55.3)[a]	(29.7)	(11.0)	
	(63.2)[b]	(24.9)	(7.9)	
B	68	23	13	104
	(59.9)	(32.2)	(12.0)	
	(63.1)	(30.4)	(10.5)	
C	58	40	12	110
	(63.3)	(34.0)	(12.7)	
	(60.7)	(35.7)	(13.6)	
D	53	38	16	107
	(61.6)	(33.1)	(12.3)	
	(53.1)	(37.9)	(16.0)	

[a]For independence model $\hat{m}_{ij}$.
[b]For uniform association model $\hat{m}_{ij}$.

with $df = 1$. The logit independence model [i.e., (7.3) with $\beta = 0$] is equivalent to the loglinear independence model (2.21), and it has estimated expected frequencies $\{\hat{m}_{ij} = n_{i+} n_{+j}/n\}$.

For the dumping severity data (Table 7.1), we noted in Sections 5.1 and 2.3 that $G^2(I) = 10.88$. Since the logit uniform association model seems to fit well, the test of H_0: $\beta = 0$ is based on $G^2(I \mid U) = 10.88 - 4.27 = 6.61$, with $\mathrm{df} = 1$. So, like the analogous (but nonequivalent) statistic (5.4) in the ordinal loglinear analysis, this test statistic gives moderately strong evidence of an association.

7.2. LOGIT MODEL FOR ORDINAL-NOMINAL TABLES

Similar types of models can be formulated that treat the levels of the explanatory variable X as nominal. In these models the association terms that are added to the basic independence model take the form of row effects. We now consider a cumulative logit model having row effects, each of which is identical for the $c - 1$ ways of forming the logits. This model is a logit analog of the loglinear row effects model. It is also analogous to the one-way analysis of variance model for continuous response variables.

Row Effects Model

The independence model is expressed in terms of the cumulative logits as

$$L_{j(i)} = \alpha_j, \qquad 1 \le i \le r, \qquad 1 \le j \le c - 1$$

With the addition of row effects for the levels of the nominal variable, this becomes

$$L_{j(i)} = \alpha_j + \tau_i, \qquad 1 \le i \le r, \qquad 1 \le j \le c - 1 \tag{7.6}$$

where $\sum \tau_i = 0$.

In model (7.6) the ith row effect is assumed to be the same for all $c - 1$ ways of forming the cumulative logits; that is, it is τ_i, not τ_{ij}. In the model there are $r(c - 1)$ logits and $(c - 1) + (r - 1)$ independent parameters, so the residual df $= (r - 1)(c - 2)$. This is the same as for the loglinear model we considered for this setting, the row effects model (5.5), but the two models are not equivalent unless $c = 2$. We refer to model (7.6) as the *logit row effects model.* Among the authors who have suggested this model in various forms are Snell (1964), Williams and Grizzle (1972), Simon (1974, Formulation B), Clayton (1974), Bock (1975, pp. 544–546), and McCullagh (1980).

As in model (7.3), α_j represents the average across the r rows of the values of the jth cumulative logit. The row effect parameters specify the nature of the association. For each pair of rows a and b, the difference in logits

$$L_{j(b)} - L_{j(a)} = \tau_b - \tau_a \tag{7.7}$$

is constant for all $c - 1$ logits. Hence the log odds ratio for the 2×2 table formed by taking rows a and b of the table and collapsing the response is assumed to be constant for all $c - 1$ collapsings. If $\tau_b > \tau_a$, then the conditional Y distribution is stochastically higher in row b than in row a.

Political Ideology Example

The data in Table 7.2 on political ideology and party affiliation were analyzed using the loglinear row effects model in Section 5.2. It seems plausible here to regard ideology as an ordinal response and party affiliation as a nominal explanatory variable. Using the ML approach described in Appendix B.3, we fitted the logit row effects model to these data. The estimates of the average logits are $\hat{\alpha}_1 = 0.532$ and $\hat{\alpha}_2 = -1.325$, and the estimates of the effects of party affiliation are $\hat{\tau}_1 = -0.670$, $\hat{\tau}_2 = -0.282$, and $\hat{\tau}_3 = 0.952$. The model can also be fitted using WLS, as described in Appendix A.2.

Table 7.2. Observed Frequencies and Estimated Expected Frequencies for Independence Model and Logit Row Effects Model for Political Ideology Data

Party Affiliation	Political Ideology			Total
	Liberal	Moderate	Conservative	
Democrat	143	156	100	399
	(102.0)[a]	(161.4)	(135.6)	
	(136.4)[b]	(170.4)	(92.2)	
Independent	119	210	141	470
	(120.2)	(190.1)	(159.7)	
	(122.5)	(203.3)	(144.3)	
Republican	15	72	127	214
	(54.7)	(86.6)	(72.7)	
	(19.9)	(65.0)	(129.1)	

[a]For independence model $\hat{m}_{ij}$.
[b]For row effects model $\hat{m}_{ij}$.

The logit row effects model predicts constant differences between pairs of rows in the $c - 1 = 2$ logits. The differences are also constant predicted log odds ratios for the two collapsings of each pair of rows into 2×2 tables. For example, $\hat{\tau}_3 - \hat{\tau}_1 = 1.622$ means that the odds of being classified conservative instead of moderate or liberal, and the odds of being classified conservative or moderate instead of liberal, are $\exp(1.622) = 5.06$ times higher for Republicans than for Democrats.

Conditional Test of Independence

Independence is the special case of model (7.6) in which $\tau_1 = \cdots = \tau_r = 0$. For the political ideology example, Table 7.2 gives the estimated expected frequencies for the independence model (I) and the cumulative logit row effects model (R). The logit model clearly gives a much better fit. Formally, $G^2(I) = 105.66$ is based on $\text{df} = (r - 1)(c - 1) = 4$, and $G^2(R) = 4.70$ is based on $\text{df} = (r - 1)(c - 2) = 2$.

Given that the row effects model holds, a conditional test of independence can be based on

$$G^2(I \mid R) = G^2(I) - G^2(R) \tag{7.8}$$

Like the analogous statistic (5.7) for the loglinear row effects model, this test statistic has asymptotically a chi-squared distribution under H_0 with df $= r - 1$. For the political ideology data we obtain $G^2(I \mid R) = 105.66 - 4.70 = 100.96$ based on df $= 2$. There is very strong evidence of an association, as there was in the ordinal loglinear analysis of these data. The association parameter estimates indicate that for the population represented by the sample in Table 7.2, Democrats tend to be the least conservative and Republicans tend to be much more conservative than the other two affiliations.

7.3. ORDINAL LOGIT MODELS FOR HIGHER DIMENSIONS

The cumulative logit models discussed in the previous two sections can be generalized to the multidimensional case where there may be both nominal and ordinal explanatory variables. These generalized models resemble multiple regression models for continuous response variables. They are simpler to construct than the analogous loglinear models in Sections 5.3 and 5.4, since it is unnecessary to model associations among the explanatory variables. The estimation procedures described in Appendixes A.2 and B.3 can be used to fit these models. Anderson and Philips (1981) gave an application of the cumulative logit model with multiple explanatory variables to the problem of discriminant analysis with an ordinal classification.

Homogeneous, Linear Logit Effects

We illustrate the multidimensional case using a three-dimensional ($r \times c \times l$) table in which X and Y are explanatory variables and Z is an ordinal response having l categories. Within each of the $r \times c$ combinations of X and Y, there are $l - 1$ cumulative logits

$$L_{k(ij)} = \log\left[\frac{(m_{ij,k+1} + \cdots + m_{ijl})}{(m_{ij1} + \cdots + m_{ijk})}\right], \qquad k = 1, \cdots, l-1 \qquad (7.9)$$

Table 7.3, which is organized by the measurement scales of the explanatory variables, lists association terms for some simple models that have linear effects of ordinal variables. Each association effect in these models is assumed identical for the $l - 1$ ways of forming the logits. Also none of these models allows for three-factor interaction. The parameters have interpretations similar to those given in Sections 7.1 and 7.2, but in terms of partial associations.

When X and Y are nominal, for instance, Table 7.3 suggests the model

$$L_{k(ij)} = \alpha_k + \tau_i^X + \tau_j^Y \qquad (7.10)$$

where $\sum \tau_i^X = \sum \tau_j^Y = 0$. The row effects $\{\tau_i^X\}$ define a stochastic ordering of

Table 7.3. Association Terms and df for Cumulative Logit Models in Three Dimensions, with Ordinal Response Z and Linear Effects of Ordinal Explanatory Variables

Scales of Explanatory Variables		Association Terms		
X	Y	X	Y	df
Ordinal	Ordinal	$\beta^X(u_i - \bar{u})$	$\beta^Y(v_j - \bar{v})$	$rcl - rc - l - 1$
Nominal	Ordinal	τ_i^X	$\beta^Y(v_j - \bar{v})$	$rcl - rc - l - c + 1$
Nominal	Nominal	τ_i^X	τ_j^Y	$rcl - rc - r - c - l + 3$

the levels of X on the ordinal response Z. This ordering is the same at every fixed level of the covariate Y. Differences between pairs of $\{\tau_i^X\}$ represent constant log odds ratios for comparing levels of X (controlling for Y) on all $l - 1$ ways of collapsing the response Z into two categories. Similar remarks apply to the $\{\tau_j^Y\}$.

Since model (7.10) contains $rc(l - 1)$ logits and $(l - 1) + (r - 1) + (c - 1)$ linearly independent parameters, it has residual $\text{df} = rcl - rc - l - r - c + 3$, as does the analogous loglinear model (5.15) for this setting. The residual df values for chi-squared goodness-of-fit tests of the other models in Table 7.3 also appear in that table. The analogous loglinear models that have the same df values use general terms for partial associations among variables that are explanatory in the logit models. For example, the X–Y association term in models (5.11) and (5.13) must be replaced by λ_{ij}^{XY}.

Dumping Severity Example

We now reconsider the three-dimensional version of the dumping severity data, discussed previously in Sections 4.5 and 5.3. Table 4.9 shows the $4 \times 4 \times 3$ cross-classification of $O =$ operation, $H =$ hospital, and $D =$ dumping severity. The standard loglinear analysis (Section 4.5) showed only weak evidence of an O–D association, but it did not utilize the ordinal nature of those variables. The analysis using a loglinear uniform association model (Section 5.3) showed strong evidence of a positive O–D association.

We shall fit three simple cumulative logit models to these data, treating dumping severity as the response variable. Let $L_{k(ij)}$ in (7.9) represent, for operation i and hospital j, the cumulative logit when the cut point for dumping severity follows category k. The data are again assumed to have resulted from an independent multinomial sample on D at each of the $4 \times 4 = 16$ O–H combinations.

There are two logits at each of the 16 O–H combinations, a total of 32 logits. The logit model

$$L_{k(ij)} = \alpha_k, \qquad k = 1, 2 \tag{7.11}$$

states that dumping severity is jointly independent of operation and hospital. This model is equivalent to the loglinear model symbolized by (D, OH). Since it has only two parameters, its residual df = 32 − 2 = 30. The G^2 value is 31.64, which indicates that this model seems to fit quite well.

Next consider the model

$$L_{k(ij)} = \alpha_k + \beta^O(u_i - \bar{u}) \tag{7.12}$$

This model assumes a linear effect of operation on the logit of dumping severity that is the same for both logits ($k = 1, 2$) and the same for each hospital. It also assumes that dumping severity is conditionally independent of hospital for each operation. Model (7.12) has only one more parameter than the independence model (7.11), and it yields $G^2 = 25.03$ based on df = 29 when fitted with integer scores. There is a marked improvement in fit compared to the independence model, as G^2 has decreased by 31.64 − 25.03 = 6.61 based on df = 1. The association parameter estimate is $\hat{\beta}^O = 0.225$. Since model (7.12) implies that D is conditionally independent of H, the table can be collapsed over the hospital dimension without changing the O–D association if the model holds. Hence these results regarding the O–D association are the same as those obtained with model (7.3) for the marginal O–D table in Section 7.1.

Finally, consider the model

$$L_{k(ij)} = \alpha_k + \beta^O(u_i - \bar{u}) + \tau_j^H \tag{7.13}$$

where $\sum \tau_j^H = 0$. For this model the logit of dumping severity is linearly related to operation and is also related to hospital through additive effects. Like the other models discussed so far in this section, this model assumes that each association parameter is homogeneous for the two ways of forming cumulative logits, and it assumes an absence of three-factor interaction. Model (7.13) yields $G^2 = 22.48$ based on df = 26. The simpler model (7.12) is the special case of model (7.13) in which all $\tau_j^H = 0$. The improvement in fit in (7.13) relative to model (7.12) is not substantial, the reduction in G^2 equaling 2.55 based on df = 3. Hence for each operation there is little evidence that the distribution of dumping severity differs among the four hospitals.

The results of fitting the three logit models are summarized in Table 7.4. The simple model (7.12) lacking hospital effects seems to fit the data adequately. This analysis yields conclusions similar to those obtained with the loglinear analysis described in Table 5.6. A loglinear model that is analo-

Table 7.4. Analysis of Cumulative Logit Models for Dumping Severity Data

Terms for Explanatory Variables	G^2	df	Difference in G^2	Difference in df
None	31.64	30		
			6.61	1
$\beta^0(u_i - \bar{u})$	25.03	29		
			2.55	3
$\beta^0(u_i - \bar{u}) + \tau_j^H$	22.48	26		

gous to logit model (7.12) is the one having linear-by-linear association between O and D, conditional independence between H and D, and a general association term relating O and H, which are the explanatory variables in the logit model. This model is

$$\log m_{ijk} = \mu + \lambda_i^O + \lambda_j^H + \lambda_k^D + \lambda_{ij}^{OH} + \beta^{OD}(u_i - \bar{u})(w_k - \bar{w})$$

Its goodness-of-fit statistic equals $G^2 = 25.35$ based on df $= 29$ when it is fitted with integer scores. The quality of fit is very similar to that obtained with logit model (7.12), although the models are not equivalent when the number of responses exceeds two.

Interaction Models

We have observed in Chapters 6 and 7 that it is quite simple to formulate logit models because of the similarity of their structure to regression models. The same remark applies to formulating logit models that allow higher-order interactions. For example, suppose that the logit of Z is linearly related to Y, but that the slope of the relationship differs across the levels of a nominal variable X. Then the second model in Table 7.3 can be generalized to

$$L_{k(ij)} = \alpha_k + \tau_i^X + \beta^Y(v_j - \bar{v}) + \zeta_i^X(v_j - \bar{v})$$

where $\sum \tau_i^X = \sum \zeta_i^X = 0$. The slope effect of Y on the logit equals $\beta^Y + \zeta_i^X$ in level i of X. This model has residual df $= rcl - rc - l - 2r + 2$. It is analogous to the analysis of covariance model (allowing interaction) for a continuous response variable and interval and nominal explanatory variables. There is no three-factor interaction if all $\zeta_i^X = 0$. See Complement 6.5 for an example of the use of this model for the case $l = 2$.

A notable feature of the logit models discussed in this section is the assumption that the effect of each explanatory variable is the same for the different ways of forming the cumulative logits. These models may be generalized to include nonhomogeneous logit effects, which means that the effects of the explanatory variables change according to which logit is formed.

Williams and Grizzle (1972) gave an example in which a model was applied that has nonhomogeneous logit effects. Whenever possible, it is desirable to use models having homogeneous effects, since the effects of the explanatory variables are then easier to summarize and interpret. If cumulative logit models having homogeneous effects fail to fit satisfactorily, it may still be useful to try fitting similar models having another logit form, such as adjacent-categories logits or the logits for Anderson's (1984) stereotype model.

NOTES

Section 7.1

1. The covariance matrix for the parameter estimates for all models in this chapter can be estimated by $-\hat{H}^{-1}$, where H is the information matrix described in Appendix B.3. This matrix can be used in constructing confidence intervals and tests, as illustrated for loglinear models at the end of Section 5.1 and in Note 5.3. Alternative tests of independence for the cumulative logit model were suggested by McCullagh (1980, pp. 116–117).
2. Suppose that model (7.3) holds for a $2 \times c$ table. McCullagh and Nelder (1983, p. 122) showed that the local odds ratios are related to the constant local-global log odds ratio β by

$$\log \theta_{1j} = \beta(F^Y_{j+1} - F^Y_{j-1}) + o(\beta), \qquad j = 1, \cdots, c-1,$$

   where $o(\beta)/\beta \rightarrow 0$ as $\beta \rightarrow 0$. This indicates that if $c > 2$, we can expect the local log odds ratios to be smaller than the local-global log odds ratio, in absolute values. It also indicates that if $|\beta|$ is small and if the scores $\{v_j = (F^Y_{j-1} + F^Y_j)/2\}$ are used in the linear-by-linear association model (5.1), then it should also fit well and have parameter β about twice the value of β for the cumulative logit model. These average cumulative probability scores are called ridits (see Section 9.3).
3. Refer to Note 4.3. For a two-way table with prior means $\{\lambda_{ij}\}$ for $\{\pi_{ij}\}$, Bishop et al. (1975) suggested the pseudo-Bayes estimator of π_{ij},

$$\tilde{p}_{ij} = \left[\frac{n}{(n + \hat{K})}\right]p_{ij} + \left[\frac{\hat{K}}{(n + \hat{K})}\right]\lambda_{ij}$$

   where $\hat{K} = (1 - \sum\sum p_{ij}^2)/[\sum\sum (p_{ij} - \lambda_{ij})^2]$ estimates the value for which the Bayes estimator has minimum risk. Further they note that the $\{\lambda_{ij}\}$ can be based on the data. For ordinal data it seems natural to let $\{\hat{\lambda}_{ij}\}$ be the best fit of some ordinal model. If the model does not hold, then asymptotically the $\{\tilde{p}_{ij}\}$ are equivalent to the $\{p_{ij}\}$, and they are (unlike the model cell proportion estimates) consistent. The smoothed estimates $\{\tilde{p}_{ij}\}$ can be considerably better estimates of the $\{\pi_{ij}\}$ than are the $\{p_{ij}\}$, particularly if the sample size is small, if there are many cells in the table, and if the smoothing model fits well. For the data in Table 7.1, $\hat{K}/(n + \hat{K})$ equals 0.73 when the loglinear uniform association model is used for the smoothing and 0.74 when the logit uniform association model is used. In each case more weight is given to the model estimate than to the sample proportion.

COMPLEMENTS

1. For a two-way table, let $L_{j(i)}$ be the jth cumulative logit in row i.

a. Show that the independence of X and Y is equivalent to

$$L_{j(i)} = \alpha_j, \qquad 1 \le i \le r, \qquad 1 \le j \le c - 1$$

b. Suppose that X is ordinal and that model (7.3) holds. Show that the rows of X are stochastically ordered with respect to their conditional distributions on Y.

c. Suppose that X is nominal and that model (7.6) holds. Show that if $\tau_a < \tau_b$, then the conditional distribution of Y is stochastically higher in row b than in row a.

2. In some cases levels of explanatory variables are not stochastically ordered on the response because of differences in dispersion. A model for the ordinal-nominal table that has this property is

$$L_{j(i)} = \frac{(\alpha_j + \tau_i)}{\lambda_i}$$

where $\sum \tau_i = \sum \log \lambda_i = 0$. This model is a special case of a nonlinear model proposed by McCullagh (1980).

a. Show that this model is saturated unless $c \ge 4$.

b. Suppose that two rows a and b have the same location, in the sense that $\tau_a = \tau_b$. Suppose that $\lambda_a > \lambda_b$. Comment on how the conditional distribution functions $F_{j(a)}$ and $F_{j(b)}$ are ordered as α_j decreases from ∞ to $-\infty$ (i.e., as the cut point j for forming the logit increases from below the first response to above all categories).

3. Consider loglinear models (5.11) and (5.13). Replace the X–Y association terms by the general term λ_{ij}^{XY}. Show that the residual df values for testing goodness of fit of these models are then the same, respec-

Table 7.5

Contact		Low			High		
Satisfaction		Low	Medium	High	Low	Medium	High
Housing	*Influence*						
Tower blocks	Low	21	21	28	14	19	37
	Medium	34	22	36	17	23	40
	High	10	11	36	3	5	23
Apartments	Low	61	23	17	78	46	43
	Medium	43	35	40	48	45	86
	High	26	18	54	15	25	62
Atrium houses	Low	13	9	10	20	23	20
	Medium	8	8	12	10	22	24
	High	6	7	9	7	10	21
Terraced houses	Low	18	6	7	57	23	13
	Medium	15	13	13	31	21	13
	High	7	5	11	5	6	13

Source: Madsen (1976). Reprinted by permission.

tively, as for the logit model in Table 7.3 in which X and Y are ordinal and the logit model in which X is nominal and Y is ordinal.

4. Table 7.5, a $4 \times 2 \times 3 \times 3$ table, is based on a sample described by Madsen (1976) of 1681 residents of twelve areas in Copenhagen. It describes interrelationships among type of housing (H), degree of contact with other residents (C), feeling of influence on apartment management (I), and satisfaction with housing conditions (S).

a. Let $L_{l(ijk)}$ be the lth cumulative logit on S at level i of H, j of C, and k of I. For the model

$$L_{l(ijk)} = \alpha_l + \tau_i^H + \tau_j^C + \beta^I(w_k - \bar{w})$$

$G^2 = 48.4$ based on $\text{df} = 41$ when integer scores are used for the levels of I. Interpret the effects on satisfaction indicated by the parameter estimates $\hat{\tau}_1^H = 0.505$, $\hat{\tau}_2^H = -0.064$, $\hat{\tau}_3^H = 0.144$, $\hat{\tau}_4^H = -0.584$, $\hat{\tau}_1^C = -0.179$, $\hat{\tau}_2^C = -0.179$, $\hat{\beta}^I = 0.635$.

b. Consider the analogous loglinear model

$$\log m_{ijkl} = \mu + \lambda_i^H + \lambda_j^C + \lambda_k^I + \lambda_l^S + \lambda_{ij}^{HC} + \lambda_{ik}^{HI} + \lambda_{jk}^{CI} + \lambda_{ijk}^{HCI} + \tau_i^{HS}(x_l - \bar{x}) + \tau_j^{CS}(x_l - \bar{x}) + \beta^{IS}(w_k - \bar{w})(x_l - \bar{x})$$

Table 7.6

Parent's Socioeconomic Status	Mental Health Status: Well	Mild Symptom Formation	Moderate Symptom Formation	Impaired
A (high)	64 (48.5)[a] (65.3, 65.9)[b]	94 (95.0) (104.4, 103.7)	58 (57.1) (50.2, 48.9)	46 (61.4) (42.1, 43.5)
B	57 (45.3) (54.2, 54.3)	94 (88.9) (94.9, 94.8)	54 (53.4) (49.9, 49.3)	40 (57.4) (45.9, 46.6)
C	57 (53.1) (55.9, 55.7)	105 (104.1) (107.2, 107.3)	65 (62.6) (61.7, 61.6)	60 (67.3) (62.2, 62.4)
D	72 (71.0) (65.3, 65.0)	141 (139.3) (137.0, 137.3)	77 (83.7) (86.4, 86.8)	94 (90.0) (95.3, 94.9)
E	36 (49.0) (39.0, 39.0)	97 (96.1) (89.6, 89.6)	54 (57.8) (61.8, 62.4)	78 (62.1) (74.7, 74.0)
F (low)	21 (40.1) (27.3, 27.7)	71 (78.7) (68.8, 68.6)	54 (47.3) (52.0, 52.5)	71 (50.9) (68.8, 68.2)

Source: Srole (1978). Reprinted by permission.
[a]For independence model $\hat{m}_{ij}$.
[b]For loglinear model $\hat{m}_{ij}$, followed by $\hat{m}_{ij}$ for logit model.

When integer scores $\{w_k = k\}$ and $\{x_l = l\}$ are used for the levels of influence and satisfaction, $G^2 = 49.2$ based on df = 41, and $\hat{\tau}_1^{HS} = 0.323$, $\hat{\tau}_2^{HS} = -0.044$, $\hat{\tau}_3^{HS} = 0.102$, $\hat{\tau}_4^{HS} = -0.381$, $\hat{\tau}_1^{CS} = -0.119$, $\hat{\tau}_2^{CS} = 0.119$, $\hat{\beta}^{IS} = 0.407$. Interpret the associations of H, C, and I with satisfaction that are indicated by these parameter estimates. Compare results to those obtained in (a) with the logit model.

5. In Table 7.6 taken from the article by Goodman (1979a) and the book by Srole et al. (1978, p. 289), the parenthesized values are estimated expected frequencies for the independence model, the loglinear uniform association model, and the cumulative logit uniform association model. The G^2 values for the three models are 47.4, 9.9, and 10.9.

a. Find the association parameter estimate for the loglinear uniform association model, and interpret its value. (*Hint:* The estimate can be obtained using the relevant expected frequency estimates.)

b. Find the association parameter estimate for the logit uniform association model, and interpret its value.

c. Find the degrees of freedom for the three models, and perform conditional tests of independence that reflect the ordinal nature of the variables. Notice the great reduction in G^2 at the loss of only one degree of freedom.

d. Compare the results of the loglinear and logit analyses in terms of the substantive information they convey about the mental health data.

6. Table 7.7, taken from Holmes and Williams (1954), refers to the relationship between tonsil size and whether a child is a carrier of the virus *Streptococcus pyogenes.* Estimated expected frequencies are given in parentheses for the independence model, the loglinear row effects model, and the cumulative logit row effects model. The G^2 values are 7.32 for the independence model, 0.39 for the loglinear model, and 0.30 for the logit model.

a. Give the results of goodness-of-fit tests for the independence, loglinear, and logit models. Use the G^2 statistics to perform conditional tests of independence that reflect the ordinal nature of tonsil size.

Table 7.7

	Tonsil Size		
	Not Enlarged	Enlarged	Greatly Enlarged
Carriers	19	29	24
	(26.6)[a]	(30.3)	(15.1)
	(18.0)[b]	(31.0)	(23.0)
	(17.8)[c]	(31.2)	(22.9)
Noncarriers	497	560	269
	(489.4)	(558.7)	(277.9)
	(498.0)	(558.0)	(270.0)
	(498.0)	(557.8)	(270.2)

Source: Holmes and Williams (1954). Reprinted by permission.

[a]For independence model $\hat{m}_{ij}$.

[b]For loglinear model $\hat{m}_{ij}$.

[c]For logit model $\hat{m}_{ij}$.

b. Find the $\hat{\tau}_1$ estimate for the loglinear row effects model, and interpret its value. (*Hint:* The estimate can be calculated using the relevant expected frequency estimates.)

c. Find the $\hat{\tau}_1$ estimate for the cumulative logit row effects model, and interpret its value.

d. Compare the results of the loglinear and logit analyses in terms of the substantive information they convey about the tonsil size data. (*Note:* Fienberg analyzed these data using continuation-ratio logits in his discussion to the paper by McCullagh 1980, p. 136.)

7. The model $\log[\pi/(1-\pi)] = \alpha + \beta x$ is often used to describe the relationship between the dose (or log dose) x of a drug that is administered and the probability $\pi = \pi(x)$ of a positive response. Suppose that we have results of a dose-response experiment at c dosage levels $x_1 < \cdots < x_c$. Then the frequencies of positive and negative responses at these doses may be displayed in a $2 \times c$ cross-classification table. If the $\{x_i\}$ are equally spaced, indicate which of the following models include this one as a special case, and explain why.

a. Loglinear independence model (2.21).

b. Loglinear row effects model (5.5).

c. Loglinear uniform association model (5.1).

d. Logit regression model (6.3).

e. Logit row effects model (7.6).

f. Logit uniform association model (7.3), but with row variable the response variable.

8. For a two-way table with column response variable that is ordinal, let $L_{j(i)} = \log(m_{i,j+1}/m_{ij}), j = 1, \cdots, c-1$. Consider the model

$$L_{j(i)} = \alpha_j + \tau_i, \quad 1 \le i \le r, \quad 1 \le j \le c-1$$

where $\sum \tau_i = 0$, referred to as the *parallel odds* model by Goodman (1983).

a. Show that the independence model is equivalent to $\tau_1 = \cdots = \tau_r = 0$.

b. Consider the loglinear row effects model (5.5) having integer column scores $\{v_j = j\}$. Show that that model is equivalent to this logit model, with the $\{\tau_i\}$ row effect parameters being identical in the two models.

c. Suppose that the row variable is also ordinal, with assigned scores $\{u_i\}$. Consider the simpler logit model for adjacent categories,

$$L_{j(i)} = \alpha_j + \beta(u_i - \bar{u}), \qquad 1 \le i \le r, \qquad 1 \le j \le c-1$$

For this model, $\{L_{j(i)},\ 1 \leq i \leq r\}$ plotted against $\{u_i,\ 1 \leq i \leq r\}$ gives parallel straight lines for $1 \leq j \leq c - 1$ having uniform slope β. Show that this model is equivalent to the linear-by-linear association model (5.1) having integer column scores $\{v_j = j\}$, with association parameter β being identical in the two models. (*Note:* These logit models for adjacent categories can be fitted using the LOG-LINEAR procedure in the SPSS^X computer package.)

CHAPTER 8

Other Models for Ordinal Variables

The preceding five chapters have been devoted solely to the loglinear and logit forms for models. This chapter describes other types of models that can be applied to ordinal categorical data. These models focus on characteristics of the association not considered in the logit and loglinear models.

The model type described in Section 8.1 has the same appearance as the ordinal loglinear models described in Chapter 5, except that scores for ordinal variables are treated as parameters. It is unnecessary for the user to assign the scores, since the estimation process provides estimated scores that yield the best fit for models such as linear-by-linear association. Section 8.2 describes another type of uniform association model, one for global rather than local or local-global odds ratios. Section 8.3 considers a simple model for the mean of a response variable. This model is probably the easiest one to interpret of the models discussed in this book, since it has exactly the same form as standard regression and analysis of variance models for interval variables. The "log-log" model discussed in Section 8.4 is likely to fit better than logit or loglinear models if the response variable has a heavy-tailed skewed distribution, such as an exponential distribution. Other approaches to modeling ordinal variables are briefly summarized in Section 8.5.

8.1. LOG-MULTIPLICATIVE MODELS

An obvious disadvantage of the ordinal loglinear models of Chapter 5 is the necessity of assigning scores to the categories of ordinal variables. For many variables no obvious choice of scores exists. Yet parameter estimates and the goodness of fit of the models depends on that choice.

In this section we consider models in which the fixed scores in loglinear models are replaced by parameters. Hence, in fitting such models, one estimates the scores that give the best fit for the loglinear model. The models are called *log-multiplicative*, because the log expected frequency is a multiplicative (rather than linear) function of the model parameters.

Multiplicative Row and Column Effects

We first illustrate the log-multiplicative model for two-dimensional tables. The loglinear models with linear-by-linear association or with row effects (or with column effects) are special cases of this model. The ordered scores $\{u_i\}$ and $\{v_j\}$ that appeared in the loglinear models are treated as unknown parameters $\{\mu_i\}$ and $\{\nu_j\}$.

The two-dimensional log-multiplicative model is

$$\log m_{ij} = \mu + \lambda_i^X + \lambda_j^Y + \beta\mu_i\nu_j \tag{8.1}$$

where $\sum \lambda_i^X = \sum \lambda_j^Y = 0$. The basic form of the model is unchanged when the $\{\mu_i\}$ or the $\{\nu_j\}$ are replaced by linear functions of themselves. Without loss of generality, therefore, an arbitrary location and scale may be assumed for them. For example, one could assume that

$$\sum \mu_i = \sum \nu_j = 0 \quad \text{and} \quad \sum \mu_i^2 = \sum \nu_j^2 = 1$$

or alternatively that

$$\sum \mu_i \pi_{i+} = \sum \nu_j \pi_{+j} = 0 \quad \text{and} \quad \sum \mu_i^2 \pi_{i+} = \sum \nu_j^2 \pi_{+j} = 1 \tag{8.2}$$

The latter constraints say that the two sets of parameters have means of zero and standard deviations of one with respect to the marginal distributions.

Due to the constraints on the model parameters, $r - 2$ of the $\{\mu_i\}$ and $c - 2$ of the $\{\nu_j\}$ are linearly independent. For testing goodness of fit, therefore

$$\text{df} = rc - [1 + (r-1) + (c-1) + 1 + (r-2) + (c-2)] = (r-2)(c-2)$$

and the table must have dimensions at least 3×3 for model (8.1) to be unsaturated. This model has been discussed by Andersen (1980, p. 211), by Clogg (1982a, 1982b), and by Goodman (1979a, 1981a, 1981b), who referred to it as "Model II" or the "*RC* model." Goodman (1981b) pointed out that this model for discrete variables has form similar to the bivariate normal density for continuous variables. When the standardized scores (8.2) are used in model (8.1), the association parameter β corresponds to $\rho/(1-\rho^2)$ in the normal density, where ρ is the Pearson correlation. A similar comment applies to the linear-by-linear association model (5.1). See Complement 8.1. Anderson (1984) proposed a logit model that is related to (8.1) when an ordering constraint is placed on the score parameters of an ordinal response variable.

If $\beta = 0$, model (8.1) simplifies to the independence model. Notice the resemblance between model (8.1) and the linear-by-linear association model (5.1). Model (8.1) can be interpreted like that model through the log odds ratios (5.3) if the fixed scores $\{u_i\}$ and $\{v_j\}$ are replaced by the parameters $\{\mu_i\}$ and $\{\nu_j\}$. This model differs from the linear-by-linear association model, though, in being invariant to interchanges of rows and columns. If model (8.1) holds and we interchange rows a and b, for example, then the model still holds with μ_a and μ_b switching places. Hence the $\{\mu_i\}$ and $\{\nu_j\}$ need not be monotonic. If this model fits well and produces parameter score estimates that are monotonic, then the linear-by-linear association model would also fit well if the fixed scores that were chosen for that model had similar spacings.

The invariance of results to orderings of categories implies that variables are treated as nominal by this model. The model can be used, however, to describe ordinal characteristics of the data. For instance, since the local log odds ratio equals

$$\log \theta_{ij} = \beta(\mu_{i+1} - \mu_i)(\nu_{j+i} - \nu_j)$$

monotonicity in the scores indicates that all local associations have the same sign. Lack of monotonicity in the scores indicates nonmonotonic associations, in the sense that local associations are positive in some locations and negative in other locations. The model can also be used for scaling purposes, when the category ordering of a hypothetically ordinal variable is not completely known. Suppose that it is plausible to assume an underlying bivariate normal distribution for the association between that variable and a second one. Then the estimated scores from the fit of the log-multiplicative model indicate a tentative ordering of the categories. See Clogg (1982a) for details.

The log-multiplicative model also resembles the row effects model (5.5) if we treat the $\{\mu_i\}$ as row effects and the $\{\nu_j\}$ as parameter scores. If μ_i and μ_{i+1} are equal, then rows i and $i+1$ have the same conditional distributions on Y, and the local log odds ratios $\{\log \theta_{ij}, 1 \leq j \leq c-1\}$ are all zero. Generally, in model (8.1) the $\{\mu_i\}$ can be regarded as row effects and the $\{\nu_j\}$ can be regarded as column effects. We will symbolize the model by RC, to reflect its multiplicative row and column effects.

Suppose that Y is an ordinal variable. If the $\{\nu_j\}$ are ordered (say $\nu_1 < \cdots < \nu_c$), then each pair of rows is stochastically ordered with respect to the column variable Y. Specifically, if $\mu_a < \mu_b$, then the conditional distribution in row b is stochastically higher than the conditional distribution in row a.

Similarly, if X is an ordinal variable, and if the $\{\mu_i\}$ are ordered, the columns can be stochastically ordered (according to the $\{\nu_j\}$) with respect to their conditional distributions on the row variable.

Inference for Log-Multiplicative Models

Although the RC model is not loglinear in the natural parameters, iterative procedures for loglinear models can be used to fit it. If one set of parameter scores is treated as fixed, then the model has loglinear form. Each cycle of the iterative procedure consists of two steps. First, the column parameter scores are treated as fixed, and the row scores are estimated as in a loglinear row effects model. Then the estimated row effects are treated as fixed row scores, and column scores are estimated as in a column effects model. Those estimates serve as fixed column scores in the first part of the next cycle. Appendixes D.1 and D.2 discuss the use of the computer packages GLIM and SPSS^X (LOGLINEAR) to fit the model. It can be more easily fitted using the program ANOAS discussed in Appendix D.5.

Haberman (1981) presented the asymptotic theory for testing H_0: $\beta = 0$ in the log-multiplicative model. The test statistic $G^2(I) - G^2(RC)$ does not have an asymptotic chi-squared distribution in this case, because the $\{\mu_i\}$ and $\{\nu_j\}$ are undetermined in the RC model if independence holds. Instead, its asymptotic distribution is the same as that of a test statistic based on the canonical correlation R_1 for the table. The canonical correlation is the maximum correlation that is obtained between the variables, out of all ways of assigning scores to the rows and to the columns (see Note 9.3).

Under the null hypothesis of independence, Haberman showed that $G^2(I) - G^2(RC)$ is asymptotically equivalent to nR_1^2, in the sense that the difference between the two converges in probability to zero. The null asymptotic distribution of either statistic is the same as that of the maximum eigenvalue of the $(r-1)$ by $(r-1)$ central Wishart matrix with $(c-1)$ degrees of freedom. The upper 1 percent and 5 percent critical values for this test are given in Table 51 of Pearson and Hartley (1972). If $r = 2$ or $c = 2$ the RC model is saturated so that $G^2(RC) = 0$, $G^2(I) - G^2(RC) = G^2(I)$, and this test reduces to the standard two-way test of independence.

Goodman (1981a) pointed out many resemblances between the RC model and the canonical correlation model, which is

$$\pi_{ij} = \pi_{i+}\pi_{+j}(1 + \lambda\mu_i\nu_j)$$

In the latter model, the scores $\{\mu_i\}$ and $\{\nu_j\}$ (also assumed to satisfy constraints such as $\sum \pi_{i+}\mu_i = \sum \pi_{+j}\nu_j = 0$ and $\sum \pi_{i+}\mu_i^2 = \sum \pi_{+j}\nu_j^2 = 1$) are estimated that produce the canonical (maximum) correlation for the joint distribution $\{\pi_{ij}\}$. That correlation is simply $\sum\sum \mu_i\nu_j\pi_{ij} = \lambda$. Goodman noted that the parameter score estimates obtained for the RC model are often very close to the score estimates obtained for the canonical correlation model.

Dumping Severity and Political Ideology Examples

In Section 5.1 we saw that the loglinear uniform association model [model (5.1) with integer scores] gives a good fit to the 4×3 cross-classification of operation severity and dumping severity (Table 5.1), with $G^2(U) = 4.59$ based on df = 5. The log-multiplicative model is a generalization of model (5.1), so its G^2 value cannot exceed 4.59. In fact this model also gives a reasonably good fit, with $G^2(RC) = 2.85$ based on df $= (r - 2)(c - 2) = 2$. For these data the independence model gave $G^2(I) = 10.88$. The test statistic $G^2(I) - G^2(RC) = 8.03$ can be used to test independence, given that the RC model fits. From Haberman (1981), the null distribution of this statistic is the same as that of the maximum eigenvalue of the 3×3 central Wishart matrix with df = 2. Table 51 of Pearson and Hartley (1972) indicates that $G^2(I) - G^2(RC)$ must exceed 10.74 to be significant at the 0.05 level and must exceed 14.57 to be significant at the 0.01 level. Hence for these data the attained significance level exceeds 0.05. The test based on the RC model does not give as strong evidence of association as the one based on $G^2(I) - G^2(U)$ for the loglinear or logit uniform association models, for which $P \simeq 0.01$.

The improvement in fit given by $G^2(U \mid RC) = G^2(U) - G^2(RC)$ provides a test of the null hypothesis that the $\{\mu_i\}$ and $\{v_j\}$ are linear transformations of the a priori selected scores for the loglinear model, given that the RC model holds. In this case $G^2(U \mid RC) = 4.59 - 2.85 = 1.74$ based on df $= 5 - 2 = 3$ is the statistic for testing the appropriateness of equal-interval scores. The use of these scores seems permissible, since the increase in G^2 is small when they are used instead of parameter scores. Hence the simpler uniform association model is adequate for describing the data.

Since the log-multiplicative model is invariant to interchanges of columns, it can also be used when the row or column variable is nominal. To illustrate, we also fit it to Table 5.3, the 3×3 cross-classification of party affiliation and political ideology. In Section 5.2 we obtained $G^2(R) = 2.81$ based on df = 2 for fitting the row effects model with integer scores for the levels of political ideology. In the more complex RC model the column scores are also parameters, and $G^2(RC) = 1.67$ based on df = 1. Again the reduction $G^2(R \mid RC) = G^2(R) - G^2(RC) = 1.14$ based on df = 1 is small, and we would accept the simpler row effects model having equal-interval scores.

Even though the row effects model is adequate for this example, let us consider the estimated scores for the RC model. When the scores are scaled so that $\sum \hat{\mu}_i = \sum \hat{v}_j = 0$ and $\sum \hat{\mu}_i^2 = \sum \hat{v}_j^2 = 1$, we get $\hat{v}_1 = -0.743$, $\hat{v}_2 = 0.079$, $\hat{v}_3 = 0.664$ for the levels of political ideology and $\hat{\mu}_1 = -0.545$, $\hat{\mu}_2 = -0.254$, $\hat{\mu}_3 = 0.799$ for the party affiliation row effects. The $\{\hat{v}_j\}$ are monotonic and nearly evenly spaced. This illustrates why the simpler row effects model with equal-interval scores (which equal $v_1 = -0.707$, $v_2 = 0$, $v_3 = 0.707$ when scaled so that $\sum v_i = 0$ and $\sum v_i^2 = 1$) fits nearly as well. To make the $\{\hat{\tau}_i\}$ from the row effects model comparable to the $\{\hat{\mu}_i\}$, we rescale them so that $\sum \hat{\tau}_i^2 = 1$. We then obtain $\hat{\tau}_1 = -0.549$, $\hat{\tau}_2 = -0.249$, and $\hat{\tau}_3 = 0.798$, nearly identical to the $\{\hat{\mu}_i\}$ for the RC model. Since the $\{\hat{v}_j\}$ are monotone and since $\hat{\mu}_1 < \hat{\mu}_2 < \hat{\mu}_3$ in the RC model, we again conclude that Democrats are less conservative (stochastically) than Independents, who are themselves much less conservative than Republicans.

Multidimensional Log-Multiplicative Models

There are various ways one can generalize the log-multiplicative model (8.1) so that it can be used with multidimensional cross-classifications of ordinal variables. Log-multiplicative models are obtained when parameters are substituted for some or all pairs of sets of fixed scores in the ordinal loglinear models of Sections 5.3 and 5.4. For example, a parameter-scores version of the homogeneous linear-effects model (5.11) is

$$\begin{aligned}\log m_{ijk} &= \mu + \lambda_i^X + \lambda_j^Y + \lambda_k^Z \\ &\quad + \beta^{XY}\mu_i v_j + \beta^{XZ}\mu_i \omega_k + \beta^{YZ} v_j \omega_k \qquad (8.3)\end{aligned}$$

where $\sum \lambda_i^X = \sum \lambda_j^Y = \sum \lambda_k^Z = 0$. The score parameters satisfy location and scale constraints such as $\sum \mu_i = \sum v_j = \sum \omega_k = 0$ and $\sum \mu_i^2 = \sum v_j^2 = \sum \omega_k^2 = 1$. The model has residual df $= rcl - 2(r + c + l) + 5$ and is always unsaturated. Model (8.3) has the properties

$$\log \theta_{ij(k)} = (\mu_{i+1} - \mu_i)(v_{j+1} - v_j)\beta^{XY}$$

$$\log \theta_{ijk} = 0$$

Hence interpretation of this model is also quite simple when the parameter scores are monotonic.

In fitting this model, we are estimating scores under which the homogeneous linear effects property holds. In other words, distances between estimated scores indicate what the spacings between rows, between columns, and between layers must be so that the partial odds ratio for two variables depends only on distances between levels and is the same across the levels of the third variable. Greater flexibility can be provided by letting the effects for each variable differ for each association, as in the model

$$\log m_{ijk} = \mu + \lambda_i^X + \lambda_j^Y + \lambda_k^Z + \beta^{XY}\mu_{1i}\nu_{1j} + \beta^{XZ}\mu_{2i}\omega_{1k} + \beta^{YZ}\nu_{2j}\omega_{2k} \qquad (8.4)$$

This model has residual df $= rcl - 3(r + c + l) + 11$. If a uniform association model fits poorly, but a model such as (8.3) or (8.4) fits well, the estimated scores reveal whether associations are nonmonotonic or simply stronger over certain regions than over others.

To illustrate these models, we consider the heart disease data of Table 6.1. In Section 6.2 we used logit models to assess the effects of blood pressure (P) and serum cholesterol (C) levels on coronary heart disease (H). For loglinear or log-multiplicative models it is unnecessary to identify a response variable, and the C–P association can also be modeled if we assume a full multinomial sampling scheme. The simple homogeneous uniform association model [(5.11) with integer scores] fits adequately, as indicated by $G^2 = 22.8$ with df $= 21$. The log-multiplicative models described above also fit well. Table 8.1 contains results of fitting these models. The decrease in G^2 obtained with the more complex models is only on the order of the decrease in degrees of freedom. In selecting a model, we must weigh the simplicity of structure and interpretation of model (5.11) against the more complete sample descriptions provided by (8.3) and (8.4).

According to the simple uniform association model, the data may be described by constant conditional log odds ratios of 0.53 for the C–H association, 0.44 for the P–H association, and 0.10 for the C–P association. All three association terms are needed in the model, since their respective standard errors are 0.11, 0.12, and 0.03. The only cell where the model fits poorly is the one having a frequency of 209, the estimated expected frequency being 183.2. This seems to be due to the restriction to uniformity of the C–P association, since the fitted value is 205.1 for the model with general C–P association term (i.e., the logit model described in Section 6.2). To make the association parameter estimates for the uniform association model comparable in value to those in the log-multiplicative models of similar form, one can rescale the fixed scores to satisfy the constraints $\sum u_i^2 = \sum v_j^2 = \sum w_k^2 = 1$. The association parameter estimates for the uniform association model then equal $\hat{\beta}^{CH} = 0.84$, $\hat{\beta}^{PH} = 0.70$, and $\hat{\beta}^{CP} = 0.48$, very similar to those for models (8.3) and (8.4).

Although the departures from uniform associations suggested by the other models are not significant, we will discuss them to illustrate the interpretation of those models. Let (i, j, k) denote level i of C, level j of P, and level k of H. First, we consider model (8.4). For the P–H association, $\hat{\nu}_{11}^P$ and $\hat{\nu}_{12}^P$ are out of order but are nearly equal. This indicates that the first two levels of P had similar distributions with respect to H in the

Table 8.1. Results of Fitting Ordinal Models to Table 6.1

Goodness of fit	Homogeneous Uniform Association	Model (8.3)	Model (8.4)
G^2	22.8	18.5	13.5
df	21	17	13
Parameter estimates			
C–H association	$\hat{\beta}^{CH} = 0.53$	$\hat{\beta}^{CH} = 0.82$	$\hat{\beta}^{CH} = 0.86$
		$\hat{\mu}_1^C = -0.46$	$\hat{\mu}_{11}^C = -0.35$
		$\hat{\mu}_2^C = -0.42$	$\hat{\mu}_{12}^C = -0.51$
		$\hat{\mu}_3^C = 0.10$	$\hat{\mu}_{13}^C = 0.08$
		$\hat{\mu}_4^C = 0.78$	$\hat{\mu}_{14}^C = 0.78$
P–H association	$\hat{\beta}^{PH} = 0.44$	$\hat{\beta}^{PH} = 0.70$	$\hat{\beta}^{PH} = 0.73$
		$\hat{\nu}_1^P = -0.60$	$\hat{\nu}_{11}^P = -0.41$
		$\hat{\nu}_2^P = -0.27$	$\hat{\nu}_{12}^P = -0.48$
		$\hat{\nu}_3^P = 0.13$	$\hat{\nu}_{13}^P = 0.12$
		$\hat{\nu}_4^P = 0.74$	$\hat{\nu}_{14}^P = 0.77$
C–P association	$\hat{\beta}^{CP} = 0.10$	$\hat{\beta}^{CP} = 0.48$	$\hat{\beta}^{CP} = 0.46$
		$\hat{\mu}_1^C = -0.46$	$\hat{\mu}_{21}^C = -0.62$
		$\hat{\mu}_2^C = -0.42$	$\hat{\mu}_{22}^C = -0.34$
		$\hat{\mu}_3^C = 0.10$	$\hat{\mu}_{23}^C = 0.37$
		$\hat{\mu}_4^C = 0.78$	$\hat{\mu}_{24}^C = 0.60$
		$\hat{\nu}_1^P = -0.60$	$\hat{\nu}_{21}^P = -0.78$
		$\hat{\nu}_2^P = -0.27$	$\hat{\nu}_{22}^P = 0.11$
		$\hat{\nu}_3^P = 0.13$	$\hat{\nu}_{23}^P = 0.06$
		$\hat{\nu}_4^P = 0.74$	$\hat{\nu}_{24}^P = 0.61$

Note: The values $\hat{\omega}_{12}^H = \hat{\omega}_{22}^H = -\hat{\omega}_{11}^H = -\hat{\omega}_{21}^H = 0.707$ are determined by the constraints.

sample, since

$$\log \hat{\theta}_{(i)11} = \hat{\beta}^{PH}(\hat{\nu}_{12}^P - \hat{\nu}_{11}^P)(\hat{\omega}_{22}^H - \hat{\omega}_{21}^H)$$
$$= -0.07$$

is close to zero compared to $\log \hat{\theta}_{(i)21}$ and $\log \hat{\theta}_{(i)31}$. Similarly, the $\{\hat{\mu}_{1i}^C\}$ in model (8.4) indicate that the first two levels of C had similar distributions on H, with $\log \hat{\theta}_{1(j)1} = -0.19$ representing a slight tendency in the sample for the incidence of heart disease to be lower at the second cholesterol level than the first. The $\{\hat{v}_{2j}^P\}$ and $\{\hat{\mu}_{2i}^C\}$ in model (8.4) indicate that the middle two levels of P had similar distributions with respect to C.

In the simpler model (8.3) a variable has the same parameter scores for each of its associations. For that model the score estimates are monotonic but $\hat{\mu}_1^C$ and $\hat{\mu}_2^C$ are quite close. This indicates that the first two levels of C had similar distributions on H and similar distributions on P in the sample. The sample data suggest through models (8.3) and (8.4) that there may be some threshold below which C and P are not associated with H. The association patterns implied by the three ordinal models are further illustrated in Table 8.2, which gives their estimated local log odds ratios for each partial association.

Table 8.2. Estimated Local Log Odds Ratios for Models Fitted to Table 6.1

Log Odds Ratio	Homogeneous Uniform Association	Model (8.3)	Model (8.4)
H–C association			
$\log \hat{\theta}_{1(j)1}$	0.53	0.05	−0.19
$\log \hat{\theta}_{2(j)1}$	0.53	0.60	0.73
$\log \hat{\theta}_{3(j)1}$	0.53	0.79	0.85
H–P association			
$\log \hat{\theta}_{(i)11}$	0.44	0.32	−0.07
$\log \hat{\theta}_{(i)21}$	0.44	0.40	0.62
$\log \hat{\theta}_{(i)31}$	0.44	0.61	0.67
C–P association			
$\log \hat{\theta}_{11(k)}$	0.10	0.01	0.11
$\log \hat{\theta}_{21(k)}$	0.10	0.08	0.29
$\log \hat{\theta}_{31(k)}$	0.10	0.11	0.09
$\log \hat{\theta}_{12(k)}$	0.10	0.01	−0.01
$\log \hat{\theta}_{22(k)}$	0.10	0.10	−0.01
$\log \hat{\theta}_{32(k)}$	0.10	0.13	−0.01
$\log \hat{\theta}_{13(k)}$	0.10	0.01	0.07
$\log \hat{\theta}_{23(k)}$	0.10	0.15	0.18
$\log \hat{\theta}_{33(k)}$	0.10	0.20	0.06

8.2. GLOBAL ODDS RATIO MODELS

For cross-classifications of ordinal variables, the loglinear uniform association model assumes constancy of the local odds ratios $\{\theta_{ij}\}$ defined in (2.15). The logit uniform association model assumes constancy of the odds ratios $\{\theta'_{ij}\}$ defined in (2.16). Those odds ratios are local in the explanatory variable, since only adjacent levels of it are used. However, they are global in the response variable, since θ'_{ij} is based on a dichotomous collapsing of that variable.

A third possible uniform association model assumes constancy of the odds ratios $\{\theta''_{ij}\}$, defined in (2.18), that are global in both variables; that is, the model

$$\theta''_{ij} = \theta, \qquad 1 \le i \le r-1, \qquad 1 \le j \le c-1 \tag{8.5}$$

assumes a constant odds ratio for all $(r-1)(c-1)$ ways of collapsing the original table into a 2×2 table. This approach is similar to the loglinear approach in the sense that the variables are treated symmetrically, no distinction being made between response and explanatory variables.

One can also formulate more complex models for the $\{\theta''_{ij}\}$. For example, the model

$$\theta''_{ij} = \alpha_i \tag{8.6}$$

states that the global odds ratio depends only on the row cut point. If a model holds for the $\{\theta''_{ij}\}$, it will generally not hold if rows or columns are permuted. Unlike the logit and loglinear row effects models, therefore, model (8.6) is generally not appropriate for nominal row variables. It is also difficult to compare pairs of rows for models such as (8.5) and (8.6), in which levels are pooled with others rather than treated individually.

Clayton (1974), Wahrendorf (1980), and Anscombe (1981) have discussed the uniform association model (8.5) for global odds ratios. Plackett (1965) defined a family of joint distributions that satisfy the global uniform association assumption. Given the marginal distribution functions F_i^X and F_j^Y, when the common value of the $\{\theta''_{ij}\}$ is $\theta \ne 1$, the joint distribution function $F_{ij} = P(X \le i, Y \le j)$ is related to θ by

$$F_{ij} = \left[\frac{1 + (\theta-1)(F_i^X + F_j^Y) - \{[1 + (\theta-1)(F_i^X + F_j^Y)]^2 - 4\theta(\theta-1)F_i^X F_j^Y\}^{1/2}}{2(\theta-1)}\right] \tag{8.7}$$

Anscombe described a conditional maximum likelihood approach for fitting the global uniform association model. If the sample marginal distributions are used to estimate the true ones, then they together with an estimate for θ determine an estimate of the joint distribution function, using (8.7). That function can be used to obtain estimated expected frequencies that satisfy the global uniform association assumption. Anscombe gave an iterative method for generating a sequence of estimates of θ that converge to the one for which the likelihood is maximized.

Wahrendorf (1980) showed how to obtain a WLS estimate of θ, and he showed how to use the sample $\{\hat{\theta}''_{ij}\}$ to test the uniform association assumption. This approach is illustrated in Appendix A.4. Semenya and Koch (1980, pp. 103–118) defined more general models for the $\{\theta''_{ij}\}$ and showed how to fit them using WLS. Goodman (1981b) argued that the assumption of uniform global association is not as appropriate as the assumption of uniform local association if the underlying distribution is bivariate normal.

For the dumping severity data of Table 5.1, WLS fitting of the global uniform association model gives an estimate of 1.55 for the common odds ratio. Hence the odds that dumping is above a certain point (instead of below it) is estimated to be 1.55 times higher if the amount of stomach removal is above a certain level instead of below it. The chi-squared statistic for testing goodness of fit of the model equals 4.48. It is based on df = 5, since there are six odds ratio responses and one parameter in the model $\theta''_{ij} = \theta$. The chi-squared statistic for testing $\theta = 1$ (or $\log \theta = 0$) equals 6.31 based on df = 1. This gives a test of independence, under the assumption that the global uniform association model holds. The results and conclusions obtained here are similar to those obtained using the loglinear or logit uniform association models. See Chapter 12 for a further comparison.

8.3. MEAN RESPONSE MODELS

Some researchers who are familiar with the use of regression and ANOVA models for continuous response variables may, at first, have difficulty developing intuition for the use of loglinear or logit models. For instance, they may not find parameters based on odds measures or log transformations to be as easily interpretable as the parameters that refer to the conditional mean of the response variable in standard regression models. To obtain a regression-type model having easily interpretable parameters, we can assign scores to the levels of the response variable and use its mean as the response function.

To illustrate, consider the two-way cross-classification of ordinal variables X and Y having scores $\{u_i\}$ and $\{v_j\}$. Within level i of X, the conditional mean of Y is $M_i = \sum_j v_j m_{ij}/n_{i+}$, $i = 1, \ldots, r$. The usual linear regression model is

$$M_i = \mu + \beta(u_i - \bar{u}), \qquad i = 1, \ldots r \tag{8.8}$$

The parameter μ is the average of the conditional means, and β is the change in the conditional mean per unit change in X. There are r responses and 2 parameters, so df $= r - 2$ and $r \geq 3$ is needed to obtain an unsaturated model. Corresponding models for multidimensional tables are equally easy to construct and interpret. The model for the ordinal-nominal table is saturated, though.

Bhapkar (1968), Grizzle et al. (1969), and Williams and Grizzle (1969) presented a WLS solution for this type of response function. This solution can be obtained with computer programs designed for using WLS to fit linear models—for example, SAS (PROC FUNCAT) and GENCAT. See Appendixes A.4 and D.4, D.6.

We fitted model (8.8) to the data in Table 5.1 on the dumping severity response and operation explanatory variable, obtaining $\hat{\mu} = 1.537$ and $\hat{\beta} = 0.075$ for a WLS solution when integer scores are used. The predicted increase in the mean dumping severity is 0.075 categories for every additional 25 percent of stomach removed. The test of H_0: $\beta = 0$ yields a chi-squared statistic of 6.37 based on df = 1. The model fits adequately, as shown by a residual chi-squared of 0.23 based on df = 2. Model (8.8), like the three uniform odds ratio models fit previously to these data, gives moderately strong evidence of an association.

Models like (8.8) can also be formulated in which the response scores are suggested by the data, rather than preassigned. For example, Semenya and Koch (1979), Koch et al. (1980), and Semenya et al. (1983) gave a WLS analysis using scores (called "ridits") that are average cumulative probabilities for the marginal response distribution. The ridit scores are described in Section 9.3.

Ordinal and interval scales are both inherently quantitative, since levels of such scales differ in magnitude. For any two observations on such scales we can ask the question, Which is higher? Nominal scales are qualitative, since levels of such scales differ in quality rather than quantity. For two observations on a nominal scale, we can ask, Are they different? but we cannot ask, Which is higher? In this respect the process of ranking observations on an ordinal scale is more similar to the process of numerically measuring observations on interval scales than the process of sorting observations on nominal scales. Indeed, ordinal variables are treated like interval variables when we assign scores to their levels as we have done for many of the models in this book. Thus one can argue that models for ordinal response variables should resemble regression models for continuous (interval) response variables more than they resemble loglinear and logit models for nominal variables. Mean response models such as (8.8) have this advantage. Fitting them is a reasonable strategy if the categorical nature of the response variable is due to crude measurement of inherently continuous variables.

Unlike the loglinear, logit, log-multiplicative, and global odds ratio models, the mean response models use summary responses from which the cell probabilities are not uniquely determined. Therefore one cannot easily use mean response models to make conclusions about structural aspects such as stochastic orderings on the response. In model (8.8), for instance, knowledge of the $\{M_i\}$ is not equivalent to knowledge of the $\{\pi_{j(i)}, j = 1, \cdots, c\}$ when $c > 2$. If model (8.8) holds, therefore, $\beta = 0$ is not equivalent to independence. The loglinear and logit models directly reflect the actual discrete way the variables are measured, and special cases of those models correspond to conditions such as independence.

8.4. PROPORTIONAL HAZARDS MODELS

Suitable response functions for a model are sometimes suggested by the form of an assumed underlying response distribution. For example, let $\{F_{j(\mathbf{x})}, j = 1, \ldots, c\}$ denote the distribution function of Y when a vector of covariates $\mathbf{X}$ takes on the value $\mathbf{x}$. If it were reasonable to assume an exponential underlying distribution for Y, or an underlying distribution of a type used in survival analysis (see Cox 1972), then $\log(1 - F_{j(\mathbf{x}_1)})$ would be approximately the same constant multiple of $\log(1 - F_{j(\mathbf{x}_2)})$ for $1 \leq j \leq c - 1$. We might therefore pose the model of a constant difference between levels of $\mathbf{X}$ on the log-log scale of the complement of the distribution function. McCullagh (1979, 1980) suggested the response function $\log[-\log(1 - F_{j(\mathbf{x})})]$ for such cases, and he argued that it would be appropriate for a wide class of underlying distributions that includes the Pareto and the Weibull.

The linear model

$$\log[-\log(1 - F_{j(\mathbf{x})})] = \alpha_j + \boldsymbol{\beta}'\mathbf{x}, \qquad 1 \leq j \leq c - 1 \tag{8.9}$$

is referred to as the *proportional hazards* model. If the order of the response categories is reversed, then $\log[-\log(F_{j(\mathbf{x})})]$ is the appropriate response function. McCullagh (1980) showed how to use the Newton-Raphson method to obtain ML estimates for a class of models that includes the log-log models and cumulative logit models. The WLS approach described in Appendix A.4 can also be used to fit the proportional hazards model. Prentice and Gloeckler (1978) used this model to analyze grouped survival

Table 8.3. Family Income Distributions (Percent) in the U.S. Northeast

Year	0–3	3–5	5–7	7–10	10–12	12–15	15+
1960	6.5	8.2	11.3	23.5	15.6	12.7	22.2
1970	4.3	6.0	7.7	13.2	10.5	16.3	42.1

Source: McCullagh (1980). Reproduced here by kind permission of the *Journal of the Royal Statistical Society*, Series B.
Note: Income in thousands of 1973 dollars.

data. Farewell (1982) generalized the model to allow for variation among the sample in the values that are regarded as category boundaries for the underlying scale.

McCullagh fit the complementary log-log model

$$\log\left[-\log\left(1 - F_{j(i)}\right)\right] = \alpha_j + \tau_i, \qquad i = 1, 2, \qquad j = 1, \cdots, 6 \quad (8.10)$$

to the family income distributions given in Table 8.3. He obtained the ML estimates $\hat{\tau}_1 = -\hat{\tau}_2 = 0.28$. The fitted distribution functions $\{\hat{F}_{j(i)}\}$ satisfy

$$\log\left[-\log\left(1 - \hat{F}_{j(1)}\right)\right] - \log\left[-\log\left(1 - \hat{F}_{j(2)}\right)\right] = \hat{\tau}_1 - \hat{\tau}_2 = 0.56$$

for all j, or

$$\log\left(1 - \hat{F}_{j(1)}\right) = (\exp(0.56)) \log\left(1 - \hat{F}_{j(2)}\right) = 1.76 \log\left(1 - \hat{F}_{j(2)}\right)$$

or

$$1 - \hat{F}_{j(1)} = \left(1 - \hat{F}_{j(2)}\right)^{1.76}$$

In words, the proportion of people who earned more than a certain amount in 1960 is estimated to equal the proportion who earned more than that amount in 1970 taken to the 1.76 power.

8.5. OTHER MODEL-BUILDING APPROACHES

The models described in this book are generally the ones that we feel are likely to be most commonly used in the near future for the analysis of ordinal variables. Many other models have been suggested, however. Some of these are quite similar in certain respects to ones already considered in this book.

The cumulative logit models discussed in Chapter 7 are linear models for a certain transformation of the distribution function of the response variable. Essentially, the inverse of the logistic distribution function is used to map the (0, 1) scale of cumulative probabilities onto the $(-\infty, \infty)$ scale consisting of possible values of linear functions of explanatory variables. Other distribution functions can be used for this mapping. McCullagh (1980) used the term "link transformation" to refer to the class of monotonic increasing transformations from the (0, 1) interval onto the $(-\infty, \infty)$ interval. The Newton-Raphson routine described in his article can be used to fit such models by maximum likelihood. Burridge (1981) and Pratt (1981) showed that the log likelihood for many of these transformations is concave, so that computational algorithms usually give rapid convergence to ML estimates.

Models that use the standard normal distribution function for this transformation are called *probit* models. For a two-way ordinal-ordinal table with column response variable, for instance, the probit linear effects model is

$$\Phi^{-1}(F_{j(i)}) = \alpha_j + \beta(u_i - \bar{u}) \tag{8.11}$$

for $1 \leq i \leq r$ and $1 \leq j \leq c - 1$, where Φ is the standard normal distribution function. This model has the same form as the cumulative logit model (7.3), with normal instead of logistic distribution function. Aitchison and Silvey (1957), Gurland et al. (1960), and Bock and Jones (1968, Ch. 8) studied models of this type for ordinal categorical responses. In practice, probit models behave quite similarly to cumulative logit models due to the similarity of logistic and normal distributions.

When there is a bivariate response, it is natural to formulate models in terms of measures of association between the response variables. Let ζ denote some measure computed between two response variables or between a response and the main explanatory variable. Let $\zeta(\mathbf{x})$ denote the value of the measure when a vector of covariates takes the value $\mathbf{x}$. We could formulate a model for $\zeta(\mathbf{x})$, such as

$$\zeta(\mathbf{x}) = \alpha + \boldsymbol{\beta}'\mathbf{x} \tag{8.12}$$

For instance, $\zeta(\mathbf{x})$ could be an assumed common value of the global odds ratio at $\mathbf{x}$. Dale (1982) studied a model of this type in which each response marginal distribution is assumed to satisfy a cumulative logit model in terms of $\mathbf{x}$.

There are many summary measures of association, besides odds ratios, that have been proposed for ordinal variables. Any of these could be used for ζ in model (8.12). The next two chapters consider analyses of ordinal data using ordinal measures of association.

NOTES

Section 8.1

1. The null asymptotic distribution of $G^2(I) - G^2(RC)$ is also the same as that of the maximum eigenvalue of the $(c - 1) \times (c - 1)$ central Wishart matrix with $r - 1$ degrees of freedom. In other words, the result of the test is identical if r and c are interchanged.
2. If model (8.1) holds, then it also holds if signs are reversed for β and the $\{\mu_i\}$, or for β and the $\{\nu_j\}$, or for the $\{\mu_i\}$ and $\{\nu_j\}$. Hence the parameters are not identifiable without further restrictions. If they are monotone, we prefer the choice of signs for which they are increasing; that is, $\mu_1 \leq \mu_2 \leq \cdots \leq \mu_r$ and $\nu_1 \leq \nu_2 \leq \cdots \leq \nu_c$. If not, we arbitrarily choose the signs that make them most similar to monotone increasing.
3. In the correlation model $\pi_{ij} = \pi_{i+}\pi_{+j}(1 + \lambda\mu_i\nu_j)$, suppose that the $\{\nu_j\}$ were preselected to be fixed monotonic scores $\{v_j\}$. Then the correlation model would be analogous (but not equivalent) to a loglinear row effects model. If the $\{\mu_i\}$ and $\{\nu_j\}$ were both preselected monotonic scores $\{u_i\}$ and $\{v_j\}$, then the correlation model would be analogous to the linear-by-linear association model. Note that if the latter correlation model holds, then

$$\sum_i u_i(\pi_{ij}/\pi_{+j}) = \lambda v_j \quad \text{and} \quad \sum_j v_j(\pi_{ij}/\pi_{i+}) = \lambda u_i$$

   Hence, the mean response model (8.8) also fits for that choice of scores, as does the analogous mean response model in which the row variable is the response.

Section 8.2

4. Anscombe (1981) gave a detailed description of properties of global odds ratio models. His book contains a program, called FOURFOLD, for fitting the uniform association model using the conditional ML approach.

Section 8.3

5. Suppose that the cumulative logit model $L_{j(i)} = \alpha_j + \beta(u_i - \bar{u})$ holds for a cross-classification of two ordinal variables. The conditional distri-

bution function of Y at level i of X then has the logistic form

$$F_{j(i)} = [\exp(L_{j(i)}) + 1]^{-1}$$
$$= [1 + \exp(\alpha_j + \beta(u_i - \bar{u}))]^{-1}, \qquad j = 1, \ldots, c-1$$

The parameters $-\alpha_1, -\alpha_2, \ldots, -\alpha_{c-1}$ are sometimes treated as scores on Y that correspond to category cut points. In that case it is natural to assume an underlying distribution for Y having conditional distribution function (at level i of X) of $F_{y(i)} = \{1 + \exp[-(y - \beta(u_i - \bar{u}))]\}^{-1}$, for which the conditional mean is $\beta(u_i - \bar{u})$. Thus for the underlying continuous response variable the mean of Y is a linear function of X, and model (8.8) holds. In this sense the cumulative logit model can be regarded as a special case of the mean response model.

COMPLEMENTS

1. Consider the form of the log-multiplicative model (8.1) in which the scores are scaled to have means of zero and standard deviations of one, as in (8.2). Express the model as a probability function for the cell probabilities $\{\pi_{ij}\}$, and demonstrate the similarity of this function with the bivariate normal density having unit standard deviations. Show that the association parameter β in the log-multiplicative model corresponds to $\rho/(1-\rho^2)$ for the bivariate normal density, where ρ is the correlation coefficient. Show that the same remark applies to the linear-by-linear association model (5.1) when the scores are scaled to satisfy $\sum u_i^2 \pi_{i+} = \sum v_j^2 \pi_{+j} = 1$. See Goodman (1981a).

2. Refer to the additive effects model described in Complement 5.6. By contrast, show that for the RC model (8.1), the local log odds ratio has the multiplicative form

$$\log \theta_{ij} = \gamma_i \delta_j$$

(*Note:* Goodman 1981a also describes a model that has both additive and multiplicative row and column effects.)

3. Refer to Complement 2 and to models (8.5) and (8.6). A log-multiplicative model for the global odds ratios $\{\theta''_{ij}\}$ has the form

$$\log \theta''_{ij} = \gamma_i \delta_j$$

Is this model invariant to permutations of rows or columns? How does this compare to the behavior of the log-multiplicative model (8.1)?

4. Once scores are assigned to levels of ordinal variables, it is tempting to use standard methodology designed for interval variables. Section 8.3 discussed the fitting of standard (mean response) regression models using WLS. If *ordinary* least squares were used instead of WLS for the dumping severity data, then $\hat{\mu} = 1.537$ and $\hat{\beta} = 0.077$. For testing H_0: $\beta = 0$, the ordinary analysis gives $t = 2.51$ based on df $= n - 2 = 415$ and a P-value of $P = 0.012$. Since df is so large, t is approximately normally distributed and $t^2 = 6.30$ has an approximate chi-squared distribution with df $= 1$ under H_0. These results are nearly identical to those obtained using WLS. Discuss the assumptions that must be made in each case for the chi-squared distribution theory to be valid.

5. Consider the row effects version of the complementary log-log model,

$$\log\left[-\log\left(1 - F_{j(i)}\right)\right] = \alpha_j + \tau_i, \qquad 1 \leq i \leq r, \qquad 1 \leq j \leq c - 1$$

where $\sum \tau_i = 0$.

a. Show that the residual df for testing goodness of fit equals $(r - 1)(c - 2)$, as is also the case for the loglinear and logit row effects models.

b. Show that independence corresponds to all $\tau_i = 0$. Indicate how to get a chi-squared statistic having df $= r - 1$ to test independence, given that the model holds.

c. Show that, if this model holds, the rows are stochastically ordered on the column variable.

6. The following data result from a study of eagle nesting in Florida. The number of eaglets produced in a nest is given for four years.

a. Use the models of this chapter to describe and interpret the data.

b. The loglinear uniform association model for these data gives $G^2 = 4.91$ based on df $= 5$, with $\exp(\hat{\beta}) = 1.12$. Does this model fit better than the independence model? Does it fit more poorly than the log-multiplicative model?

c. How must we regard the samples in order for these inferences to be valid?

	Eaglets		
Year	0	1	2 or 3
1978	102	63	64
1979	106	90	83
1980	99	72	113
1981	100	92	116

Source: Steve Nesbitt, Florida Game and Fresh Water Fish Commission. Used by permission.

CHAPTER 9

Measures of Association for Ordinal Variables

In Chapter 2 we introduced some simple measures of association for cross-classification tables. The next two chapters describe specialized measures of association for tables containing ordinal variables. These measures, unlike the ones presented in Chapters 3 through 8, are not model based. They have the advantage of simplicity, being single numbers that are applicable regardless of whether the models discussed in this text fit the data. They have the disadvantage of being less informative. For instance, they do not imply certain patterns for cell expected frequencies.

The first two sections of this chapter treat the case in which both variables X and Y are ordinal. In Section 9.3 summary measures are introduced for the ordinal-nominal table. Analogous partial association measures for multidimensional tables are discussed in Section 9.4. In the final section we consider the effects of the choice of categories for an ordinal variable on such matters as power in detecting association and bias in measuring an underlying association for continuous variables.

9.1. CONCORDANCE AND DISCORDANCE

The most commonly used measures of association for ordinal variables are those based on the numbers of concordant and discordant pairs of observations in the sample. A pair of observations is *concordant* if the member that ranks higher on variable X also ranks higher on variable Y. A pair of observations is *discordant* if the member that ranks higher on X ranks lower on Y.

We shall illustrate these definitions using Table 9.1, taken from the 1972 General Social Survey of the National Data Program. The ordinal variables

Table 9.1. Cross-Classification of Attitude Toward Abortion by Amount of Schooling

	Attitude toward Abortion			
Schooling	Generally Disapprove	Middle Position	Generally Approve	Total
Less than high school	209	101	237	547
High school	151	126	426	703
More than high school	16	21	138	175
Total	376	248	801	1425

are X = amount of schooling, with levels less than high school $(<HS)$, high school (HS), and more than high school $(>HS)$, and Y = attitude toward abortion, with levels generally approve (A), middle position (M), and generally disapprove (D). Level A will be treated as the high end of the scale on Y.

Consider a pair of individuals, one of whom is classified in the cell $(<HS, D)$ on (X, Y) and the other in the cell (HS, M). This pair of individuals is concordant, since the second individual is ranked higher than the first both in amount of schooling (HS vs. $<HS$) and in attitude toward abortion (M vs. D). Now each of the 209 individuals classified in the cell $(<HS, D)$ form concordant pairs when matched with each of the 126 individuals classified (HS, M), so there are $209 \times 126 = 26{,}334$ concordant pairs from these two cells. Similarly, the 209 individuals in the cell $(<HS, D)$ are part of a concordant pair when matched with each of the 426 individuals classified (HS, A) and the $21 + 138$ individuals classified higher on both variables. Using the same reasoning, the 101 individuals in the cell $(<HS, M)$ are part of concordant pairs when matched with the $426 + 138$ individuals ranked higher on both variables.

Figure 9.1 illustrates the pairings of cells that result in concordant pairs. The total number of concordant pairs in this sample, denoted by C, equals

$$\begin{aligned} C &= 209(126 + 426 + 21 + 138) + 101(426 + 138) \\ &\quad + 151(21 + 138) + 126(138) \\ &= 246{,}960 \end{aligned}$$

The general formula for calculating C is

$$C = \sum_{i<k} \sum_{j<l} n_{ij} n_{kl} \tag{9.1}$$

where the first summation is over all pairs of rows $i < k$ and the second summation is over all pairs of columns $j < l$. The number of discordant

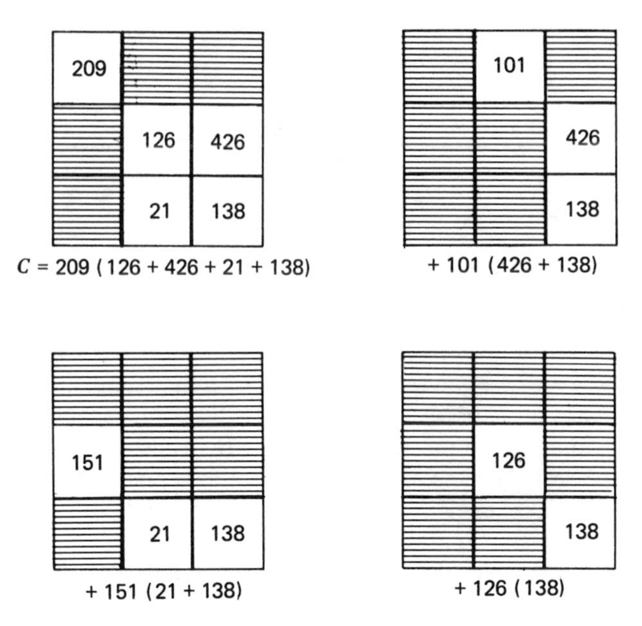

Figure 9.1. Illustration of concordant pairs.

pairs of observations is

$$D = \sum_{i<k} \sum_{j>l} n_{ij} n_{kl} \tag{9.2}$$

which for Table 3.1 equals

$$\begin{aligned} D &= 237(151 + 126 + 16 + 21) + 101(151 + 16) \\ &\quad + 426(16 + 21) + 126(16) \\ &= 109{,}063 \end{aligned}$$

The greater the relative number of concordant pairs, the more evidence there is of a positive association, in the sense that low Y's tend to occur with low X's and high Y's with high X's. In this example $C > D$ suggests a tendency for individuals having greater schooling to have a more favorable attitude toward abortion.

Not all $\binom{n}{2} = n(n-1)/2$ pairs of observations are concordant or discordant. In Table 9.1 there are $1425 \times 1424/2 = 1{,}014{,}600$ pairs, nearly triple the total number of concordant pairs plus discordant pairs. These additional pairs are ones that are *tied* on one or both of the variables. For example, 547 individuals were classified $<HS$ on schooling, so $547 \times 546/2$ pairs were tied at that level of the row variable. Summing over the three

levels of X = schooling, we see that the total number of pairs tied on that variable is

$$T_X = [547 \times 546 + 703 \times 702 + 175 \times 174]/2 = 411{,}309$$

Generally, the total number of pairs tied on the row variable X is

$$T_X = \sum \frac{n_{i+}(n_{i+} - 1)}{2}$$

Similarly, the number of pairs tied on the column variable Y is

$$T_Y = \sum \frac{n_{+j}(n_{+j} - 1)}{2}$$

which for Table 9.1 equals 421,528.

Some of the pairs tied on X are also tied on Y, namely pairs of observations from the same cell. The total number of these pairs is

$$T_{XY} = \sum_i \sum_j \frac{n_{ij}(n_{ij} - 1)}{2}$$

which for Table 9.1 equals 174,260. The total number of pairs can be expressed as

$$\frac{n(n-1)}{2} = C + D + T_X + T_Y - T_{XY} \tag{9.3}$$

In this formula T_{XY} is subtracted because pairs tied on both X and Y have been counted twice, once in T_X and once in T_Y.

9.2. ORDINAL-ORDINAL MEASURES OF ASSOCIATION

We now introduce some measures of association for cross-classifications of ordinal variables. Several measures are based on the difference $C - D$ between the numbers of concordant and discordant pairs. For each of these measures the association is said to be positive if $C - D > 0$ and negative if $C - D < 0$. In each case a reversal in the category orderings of one variable simply causes a change in the sign of the measure.

Gamma

Of the $C + D$ pairs of observations that are untied on both variables, $C/(C + D)$ is the proportion of concordant pairs and $D/(C + D)$ is the proportion of discordant pairs. The difference between these two proportions is the measure *gamma* proposed by Goodman and Kruskal (1954),

$$\hat{\gamma} = \frac{(C - D)}{(C + D)} \tag{9.4}$$

The population version of gamma is

$$\gamma = \frac{(\Pi_c - \Pi_d)}{(\Pi_c + \Pi_d)}$$

where

$$\Pi_c = 2\sum_{i<k}\sum_{j<l} \pi_{ij}\pi_{kl} \quad \text{and} \quad \Pi_d = 2\sum_{i<k}\sum_{j>l} \pi_{ij}\pi_{kl}$$

are the probabilities of concordance and discordance for a randomly selected pair of observations. The factor of 2 occurs in these formulas because the first observation could be in cell (i, j) and the second in cell (k, l), or vice versa.

The properties of γ (and $\hat{\gamma}$) follow directly from its definition. Its range of values is $-1 \leq \gamma \leq 1$, with $\gamma = 1$ if $\Pi_d = 0$ and $\gamma = -1$ if $\Pi_c = 0$. Table 9.2 illustrates cross-classifications having various values of γ. Note that $|\gamma| = 1$ implies that the relationship is monotone but not necessarily strictly monotone. If $\gamma = 1$, in other words, then for a pair of observations (X_a, Y_a) and (X_b, Y_b) with $X_a < X_b$, it follows that $Y_a \leq Y_b$ but not necessarily that $Y_a < Y_b$. Independence implies that $\gamma = 0$, but the converse of this does not hold. To illustrate, Table 9.2*c* shows a U-shaped bivariate distribution for which $\Pi_c = \Pi_d$ and hence $\gamma = 0$. The value of γ is symmetric in the sense that it is the same whether X or Y (or neither) is regarded as the response variable.

Table 9.2. Values of Gamma for Various Cross-Classifications

a. $\gamma = 1$	$\frac{1}{3}$	0	0
	0	$\frac{1}{3}$	0
	0	0	$\frac{1}{3}$
b. $\gamma = 1$	0.2	0	0
	0.2	0.2	0
	0	0.2	0.2
c. $\gamma = 0$	0.2	0	0.2
	0.2	0	0.2
	0	0.2	0
d. $\gamma = -1$	0	0.30	
	0.03	0.67	

For the case of 2 × 2 tables, γ simplifies to

$$\gamma = \frac{(\pi_{11}\pi_{22} - \pi_{12}\pi_{21})}{(\pi_{11}\pi_{22} + \pi_{12}\pi_{21})}$$

This measure is also referred to as Yule's Q and is related to the odds ratio $\theta = \pi_{11}\pi_{22}/\pi_{12}\pi_{21}$ by $\gamma = (\theta - 1)/(\theta + 1)$. Thus gamma is a monotonic function of θ that transforms from a $[0, \infty]$ scale to a $[-1, +1]$ scale.

Somers' *d*

Somers (1962) proposed a measure similar to gamma, but for which pairs untied on X serve as the base rather than only those untied on both X and Y. The sample version

$$d_{YX} = \frac{(C - D)}{\{n(n-1)/2 - T_X\}} \tag{9.5}$$

is the difference between the proportions of concordant and discordant pairs, out of the pairs that are untied on X. This is an asymmetric measure intended for use when Y is a response variable. The symbol d_{XY} denotes the version that has $\{n(n-1)/2 - T_Y\}$ in the denominator.

Since $n(n-1)/2 - T_X = C + D + (T_Y - T_{XY}) \geq C + D$, the denominator of d_{YX} is at least as large as the denominator of $\hat{\gamma}$, and hence $|d_{YX}| \leq |\hat{\gamma}|$. In order for $|d_{YX}| = 1$, there must be stricter monotonicity than for $|\hat{\gamma}| = 1$, in the sense that C or D must equal 0 and in addition none of the pairs that are untied on X can be tied on Y.

The population version of Somers' d is

$$\Delta_{YX} = \frac{(\Pi_c - \Pi_d)}{\left[1 - \sum_i \pi_{i+}^2\right]}$$

It can be shown that for 2 × 2 tables Δ_{YX} simplifies to the difference of proportions $\pi_{1(1)} - \pi_{1(2)} = \pi_{11}/\pi_{1+} - \pi_{21}/\pi_{2+}$.

Kendall's tau-*b*

Yet another variant of gamma was proposed by Kendall (1945). In its sample form it equals

$$\hat{\tau}_b = \frac{(C - D)}{\sqrt{[n(n-1)/2 - T_X][n(n-1)/2 - T_Y]}} \tag{9.6}$$

and is referred to as *Kendall's tau-b*. The population version of tau-b is

$$\tau_b = \frac{(\Pi_c - \Pi_d)}{\sqrt{\left[1 - \sum_i \pi_{i+}^2\right]\left[1 - \sum_j \pi_{+j}^2\right]}}$$

For 2 × 2 tables, $\hat{\tau}_b$ equals the Pearson correlation obtained when scores are assigned to the rows and to the columns. Since $C + D$ can be no greater than $n(n-1)/2 - T_X$ or $n(n-1)/2 - T_Y$, it also cannot exceed their geometric average (the denominator of $\hat{\tau}_b$), so that $|\hat{\tau}_b| \leq |\hat{\gamma}|$. Like $\hat{\gamma}$, $\hat{\tau}_b$ is a symmetric measure. It is related to the asymmetric Somers' d values by

$$\hat{\tau}_b^2 = d_{YX} d_{XY}$$

Thus $\hat{\tau}_b$ is the geometric average of d_{YX} and d_{XY}. This relationship between $\hat{\tau}_b$ and Somers' d is reminiscent of the relationship $r^2 = b_{YX} b_{XY}$ between the Pearson correlation r and the least squares slopes for regressing Y on X and X on Y. We shall see now that, in fact, $\hat{\tau}_b$ is a special case of a correlation and Somers' d is a special case of a least squares slope for a linear model defined using pair scores.

For each pair of observations (X_a, Y_a) and (X_b, Y_b), let

$$X_{ab} = \text{sign}(X_a - X_b) = \begin{cases} -1 & \text{if } X_a < X_b \\ 0 & \text{if } X_a = X_b \\ 1 & \text{if } X_a > X_b \end{cases}$$

$$Y_{ab} = \text{sign}(Y_a - Y_b) = \begin{cases} -1 & \text{if } Y_a < Y_b \\ 0 & \text{if } Y_a = Y_b \\ 1 & \text{if } Y_a > Y_b \end{cases}$$

The sign scores $\{X_{ab}\}$ are indicator variables that index whether X_a is greater than or less than X_b, and similarly for the $\{Y_{ab}\}$. Note that $X_{ab} = -X_{ba}$ and $Y_{ab} = -Y_{ba}$. The product $X_{ab} Y_{ab}$ equals 1 for a concordant pair and -1 for a discordant pair. The square X_{ab}^2 equals 1 for a pair untied on X and equals 0 for a pair tied on X. Similarly, Y_{ab}^2 equals 1 for a pair untied on Y and equals 0 for a pair tied on Y.

For the $n(n-1)$ ordered pairs of observations (a, b) with $a \neq b$,

$$\sum_{a \neq b}\sum X_{ab} Y_{ab} = 2(C - D)$$

$$\sum_{a \neq b}\sum X_{ab}^2 = 2\left[\frac{n(n-1)}{2} - T_X\right]$$

$$\sum_{a \neq b}\sum Y_{ab}^2 = 2\left[\frac{n(n-1)}{2} - T_Y\right]$$

$$\sum_{a \neq b}\sum X_{ab} = \sum_{a \neq b}\sum Y_{ab} = 0$$

Each pair is used twice in these sums so that $\sum\sum X_{ab} = \sum\sum Y_{ab} = 0$ due to the relationships $X_{ab} = -X_{ba}$ and $Y_{ab} = -Y_{ba}$. The sample correlation between the $\{X_{ab}\}$ and $\{Y_{ab}\}$ is therefore

$$\frac{\sum_{a \neq b}\sum X_{ab} Y_{ab}}{\sqrt{\left(\sum_{a \neq b}\sum X_{ab}^2\right)\left(\sum_{a \neq b}\sum Y_{ab}^2\right)}}$$

$$= \frac{(C - D)}{\sqrt{[n(n-1)/2 - T_X][n(n-1)/2 - T_Y]}} = \hat{\tau}_b$$

Similarly, the least squares slope for regressing $\{Y_{ab}\}$ on $\{X_{ab}\}$ is d_{YX}. See Hawkes (1971) and Ploch (1974) for ordinal analogs of partial slopes and correlations obtained by constructing a multiple regression model for sign scores.

Kendall's tau

The notions of concordance and discordance were first introduced for continuous variables. In that case samples can be fully ranked on both variables, so that $T_X = T_Y = T_{XY} = 0$. Therefore $C + D = n(n-1)/2$, and

$$\hat{\gamma} = d_{YX} = \hat{\tau}_b = \frac{(C - D)}{n(n-1)/2} \tag{9.7}$$

Similarly, in the population the common value of γ, Δ_{YX}, and τ_b is $\Pi_c - \Pi_d$ when the variables are continuous. This coefficient of rank correlation was proposed originally by Kendall (1938) and is called *Kendall's tau* and denoted by τ. It is the difference between the proportions of concordant and discordant pairs, out of *all* pairs of observations. Daniels (1944) noted that tau is a correlation coefficient for sign scores, and then Kendall (1945) formulated tau-*b* by constructing the same correlation for the case when some ties exist. In the discrete case a large proportion of the pairs are tied, so τ_b is normally used instead of τ since τ cannot possibly attain a very large value.

Other Correlation Coefficients

Yet other ordinal measures have been proposed that use $C - D$ as a numerator (e.g., see tau-c in Complement 9.2). In addition, other classes of ordinal measures can be expressed as correlation coefficients calculated for certain scoring systems. If equally spaced scores are assigned to the rows and to the columns, the Pearson correlation coefficient equals

$$R = \frac{\sum_i \sum_j (i - \mu_1)(j - \mu_2)\pi_{ij}}{\left\{\left[\sum_i (i - \mu_1)^2 \pi_{i+}\right]\left[\sum_j (j - \mu_2)^2 \pi_{+j}\right]\right\}^{1/2}} \tag{9.8}$$

where $\mu_1 = \sum_i i\pi_{i+}$ and $\mu_2 = \sum_j j\pi_{+j}$.

If rank-type scores are used, the correlation becomes a generalization of Spearman's rho. Let

$$r_i^X = \sum_{k=1}^{i-1} \pi_{k+} + \frac{\pi_{i+}}{2}, \qquad i = 1, \ldots, r$$

$$r_j^Y = \sum_{l=1}^{j-1} \pi_{+l} + \frac{\pi_{+j}}{2}, \qquad j = 1, \ldots, c$$

The $\{r_i^X\}$ are average cumulative probabilities for the row marginal distribution, and the $\{r_j^Y\}$ are average cumulative probabilities for the column marginal distribution. The analog of Spearman's rho considered by Kendall (1970, p. 38) can be expressed as

$$\rho_b = \frac{\sum_i \sum_j (r_i^X - 0.5)(r_j^Y - 0.5)\pi_{ij}}{\left\{\left[\sum_i (r_i^X - 0.5)^2 \pi_{i+}\right]\left[\sum_j (r_j^Y - 0.5)^2 \pi_{+j}\right]\right\}^{1/2}} \tag{9.9}$$

For 2×2 tables, $\tau_b = \rho_b = R$.

There is much literature concerning the estimation of the Pearson correlation coefficient for a continuous bivariate distribution that is assumed to underly the cross-classification table. See, for instance, Lancaster and Hamdam (1964).

Odds Ratio Measures

The measures discussed in this section summarize the association in the table by a single number. When $r > 2$ or $c > 2$, this involves a reduction in information, in the sense that the $\{n_{ij}\}$ cannot be reconstructed from the marginal frequencies and the value of the measure of association. To avoid a loss of information, we can describe the table through a set of $(r-1)(c-1)$ odds ratios, as described in Section 2.4.

Attitude on Abortion Example

To illustrate these ordinal measures, we return to Table 9.1 on schooling and attitude toward abortion. For the 1,014,600 pairs in those data, $C = 246{,}960$, $D = 109{,}063$, $T_X = 411{,}309$, $T_Y = 421{,}528$. Of the untied pairs, 69.37 percent are concordant and 30.63 percent are discordant, the difference of which is

$$\hat{\gamma} = \frac{(246{,}960 - 109{,}063)}{(246{,}960 + 109{,}063)} = 0.6937 - 0.3063 = 0.387$$

The values of Somers' d and Kendall's tau-b are

$$d_{YX} = \frac{(246{,}960 - 109{,}063)}{(1{,}014{,}600 - 411{,}309)} = 0.229 \quad \text{and}$$

$$\hat{\tau}_b = \frac{(246{,}960 - 109{,}063)}{(1{,}014{,}600 - 411{,}309)(1{,}014{,}600 - 421{,}528)} = 0.231$$

All three measures indicate that greater schooling tends to correspond to a more positive attitude on abortion.

Of the measures introduced in this section, gamma seems to be the most frequently used. It is probably the easiest of the concordance-based measures to interpret. However, in at least one important characteristic tau-b is a superior measure. The value of gamma tends to be more dependent than tau-b on the number of categories and the way they are defined. In the next section we shall see that Somers' d is a particularly useful measure for $2 \times c$ tables in which the column variable is an ordinal response variable, even if the rows are unordered.

9.3. ORDINAL-NOMINAL MEASURES OF ASSOCIATION

We now consider measures for tables in which only one of the two classifications is ordered, the other one being nominal. We first study the case in which the nominal variable is a dichotomy. For example, we may be interested in comparing two groups (men and women, whites and blacks, drug A and drug B) on a response variable having ordered categories. The frequencies are then displayed in a $2 \times c$ table.

The association can be described in terms of the difference between the two levels of the nominal variable with respect to their conditional distributions on the ordinal variable. These conditional distributions are, for $i = 1$ and 2, $\{\pi_{j(i)} = \pi_{ij}/\pi_{i+}, j = 1, \ldots, c\}$. Let Y_1 and Y_2 be independent random variables having these two distributions. In other words, Y_1 and Y_2 denote the column numbers of the ordinal variable for members selected at random from rows 1 and 2, independently of each other.

Measures Based on $P(Y_i > Y_j)$

If the distributions of Y_1 and Y_2 are not identical, it usually is important to check whether Y_1 tends to be larger than Y_2, or vice versa. The ordinal nature of Y_1 and Y_2 can be utilized through simple probabilities such as $P(Y_1 > Y_2)$ and $P(Y_2 > Y_1)$. These probabilities do not generally sum to 1, since the response variable is discrete, so it is helpful to summarize their relative sizes. Two useful comparisons are

$$\Delta = P(Y_2 > Y_1) - P(Y_1 > Y_2)$$
$$= \sum_{i<j}\sum \pi_{i(1)}\pi_{j(2)} - \sum_{i>j}\sum \pi_{i(1)}\pi_{j(2)} \qquad (9.10)$$

and

$$\alpha = \frac{P(Y_2 > Y_1)}{P(Y_1 > Y_2)}$$
$$= \frac{\sum_{i<j}\sum \pi_{i(1)}\pi_{j(2)}}{\sum_{i>j}\sum \pi_{i(1)}\pi_{j(2)}}$$
$$= \frac{\sum_{i<j}\sum \pi_{1i}\pi_{2j}}{\sum_{i>j}\sum \pi_{1i}\pi_{2j}} \qquad (9.11)$$

Like gamma, tau-*b*, and Somers' *d*, Δ falls in the range $[-1, +1]$, whereas $0 \leq \alpha \leq \infty$. Independence implies that $\Delta = 0$ and $\alpha = 1$ ($\log \alpha = 0$), but these values do not imply independence. The sample versions of these measures make sense for full multinomial sampling or for independent multinomial sampling within the rows.

Upon substituting $\pi_{j(i)} = \pi_{ij}/\pi_{i+}$, we can express Δ as

$$\Delta = \frac{\left(\sum_{i<j}\sum \pi_{1i}\pi_{2j} - \sum_{i>j}\sum \pi_{1i}\pi_{2j}\right)}{\pi_{1+}\pi_{2+}}$$

$$= \frac{\left(2 \sum_{i<j} \sum \pi_{1i}\pi_{2j} - 2 \sum_{i>j} \sum \pi_{1i}\pi_{2j}\right)}{[1 - (\pi_{1+}^2 + \pi_{2+}^2)]} \tag{9.12}$$

since $1 = \pi_{1+} + \pi_{2+} = (\pi_{1+} + \pi_{2+})^2 = \pi_{1+}^2 + \pi_{2+}^2 + 2\pi_{1+}\pi_{2+}$. Now if the rows are ordered, the numerator of the last expression in (9.12) equals $\Pi_c - \Pi_d$, and hence Δ is simply Somers' Δ_{YX}. Generally, dichotomous nominal variables can be used in methods designed for ordinal variables, since reversing the two categories changes the direction but not the magnitude of the association and does not produce different substantive conclusions.

The measure α can also be related to a previously discussed measure. For 2×2 tables, α is simply the odds ratio $\pi_{11}\pi_{22}/\pi_{12}\pi_{21}$. Thus α can be regarded as a generalized odds ratio for ordinal data. For further discussion of the properties of α, see Agresti (1980).

Ridit Analysis

Consider now an $r \times c$ table in which the rows are levels of a nominal variable and the columns are levels of an ordinal variable. A measure such as Δ can be used to compare each pair of rows. Also one can construct some weighted average of pairwise values to summarize the differences among the rows. An index of this type is the ordinal-nominal measure of association called *Freeman's theta* (see Freeman 1976, Agresti 1981a). We shall now discuss an alternative approach, whereby a score is calculated for each row that summarizes how high its observations tend to be on the ordinal scale.

If ordered scores $\{v_j\}$ are assigned to the columns, we can compute the mean response

$$M_i = \sum_j v_j \pi_{j(i)}$$

in each row. This measure is the response used in models in Section 8.3. In *ridit analysis*, introduced by Bross (1958), the researcher uses cumulative probability scores instead of arbitrarily selecting the scores for the columns. Consider the marginal distribution $\{\pi_{+j}, j = 1, \ldots, c\}$ of the response variable Y. The jth ridit r_j is the average cumulative probability within category j of the ordinal response,

$$r_j = \sum_{k=1}^{j-1} \pi_{+k} + (\tfrac{1}{2})\pi_{+j}, \qquad j = 1, \ldots, c \tag{9.13}$$

that is, the jth ridit is the proportion of individuals below category j plus half the proportion in category j.

For visual relief, we have suppressed the Y superscript in the r_j^Y notation used previously in (9.9). Note that

$$r_j = \frac{(F_{j-1}^Y + F_j^Y)}{2}$$

where $F_j^Y = \sum_{k \leqslant j} \pi_{+k}$, $j = 1, \ldots, c$, is the distribution function of Y. Suppose that the categories of the ordinal variable represent intervals of an underlying continuous distribution. If the underlying distribution were uniform over each interval, then r_j would equal the probability that a randomly selected individual falls below the midpoint of category j.

The ridits have the ordering $r_1 < r_2 < \cdots < r_c$, and they can be used like category scores. For example, the *mean ridit*

$$R_i = \sum_j r_j \pi_{j(i)} \tag{9.14}$$

can be computed at each level of the nominal variable. The mean ridit R_i has the property

$$R_i = P(Y_i > Y^*) + (\tfrac{1}{2})P(Y_i = Y^*)$$

where Y_i is the response category for a randomly selected observation in row i and Y^* is the response category for a randomly selected observation from the marginal distribution $\{\pi_{+j}, j = 1, \ldots, c\}$ of Y. In other words, R_i approximates the underlying probability that an individual from level i of the nominal variable ranks higher on the ordinal variable than does an individual in the overall population. In addition the quantity $R_i - R_k + 0.5$ approximates the probability that a randomly selected individual from row i ranks higher than a randomly selected individual from row k, for the underlying continuous distribution. These statements are only approximate, since it is unknown how the tied observations in the discrete case would be ordered for the underlying continuum. For the special case of $2 \times c$ tables the difference in mean ridits is related to the delta measure (9.10) by $\Delta = 2(R_2 - R_1)$.

The first three letters of the term "ridit" refer to the use of R_i as a measure of how one distribution (the conditional distribution of Y in row i) compares *r*elative to an *i*dentified *d*istribution (the marginal distribution of Y in this case). The mean of the ridits taken with respect to the marginal distribution used to define the ridits, $\sum_j \pi_{+j} r_j$, always equals 0.50. Also the weighted average $\sum_i \pi_{i+} R_i$ of the mean ridits always equals 0.50.

Although ridits are often used as a mechanism for scoring levels of an ordinal variable, it can be misleading to regard them as realistic values for an underlying scale having nonuniform distribution. For instance, adjacent categories that are clearly different will have nearly identical ridit scores if they both have relatively small probabilities.

Tonsil Size Example

The summary measures presented in this section make the most sense when the conditional distributions of Y within the various rows are stochastically ordered. If row 2 is stochastically higher than row 1, for instance, it can be shown that $\Delta \geq 0$, $\alpha \geq 1$, and $R_2 \geq R_1$. We illustrate these measures using the data in Table 9.3, taken from Holmes and Williams (1954). Children are classified on tonsil size and on whether they are carriers of the virus *Streptococcus pyogenes.* The sample conditional distributions are given in parentheses under the sample frequencies. The tonsil size distribution is stochastically higher for carriers than for noncarriers.

For Table 9.3 the sample values of Δ and α are

$$\hat{\Delta} = \frac{\left[\sum\sum_{i<j} n_{1i}n_{2j} - \sum\sum_{i>j} n_{1i}n_{2j}\right]}{n_{1+}n_{2+}} = 0.170 \quad \text{and}$$

$$\hat{\alpha} = \frac{\left(\sum\sum_{i<j} n_{1i}n_{2j}\right)}{\left(\sum\sum_{i>j} n_{1i}n_{2j}\right)} = 1.69$$

The $\hat{\alpha}$ value means that there are 1.69 times as many pairs of noncarriers and carriers for which the carrier has larger tonsils as there are pairs for which the noncarrier has larger tonsils.

The lower margin of Table 9.3 contains the sample marginal distribution on tonsil size. From this distribution the sample ridits are $\hat{r}_1 = 0.369/2 = 0.185$, $\hat{r}_2 = 0.369 + 0.421/2 = 0.580$, and $\hat{r}_3 = 0.369 + 0.421 + 0.210/2 =$

Table 9.3. Tonsil Size by Whether Carrier of *Streptococcus pyogenes*

	Tonsil Size			
	Not Enlarged	Enlarged	Greatly Enlarged	Total
Noncarriers	497 (0.375)	560 (0.422)	269 (0.203)	1326
Carriers	19 (0.264)	29 (0.403)	24 (0.333)	72
Total	516 (0.369)	589 (0.421)	293 (0.210)	1398

Note: Distributions on tonsil size in parentheses.

0.895. The mean ridit for carriers is

$$\hat{R}_2 = \sum_j \hat{r}_j p_{j(2)} = 0.581$$

This gives an estimate of the probability that a randomly selected carrier has larger tonsils than an individual randomly selected from the population represented by the entire sample. The mean ridit for noncarriers is $\hat{R}_1 = 0.496$, necessarily close to 0.5 since noncarriers comprise nearly all of the sample.

For the underlying continuous distribution of tonsil size, the estimated probability that a carrier has larger tonsils than a noncarrier is $\hat{R}_2 - \hat{R}_1 + 0.5 = 0.585$. This number is the same as for the observed discrete sample,

$$\hat{P}(Y_2 > Y_1) + \left(\frac{1}{2}\right)\hat{P}(Y_1 = Y_2) = \sum_{i<j}\sum p_{i(1)} p_{j(2)} + \sum_i p_{i(1)} p_{i(2)}$$

$$= \frac{(\hat{\Delta} + 1)}{2}$$

Hence the analyses using the $\hat{\Delta}$ (or $\hat{\alpha}$) measures of association and using the mean ridits give us the same conclusions regarding the tendency for carriers to have larger tonsils than noncarriers for this sample. For $r \times c$ tables with $r > 2$, $\hat{R}_i - \hat{R}_k + 0.5$ is not identical to $(\hat{\Delta} + 1)/2$ computed for those rows, though these measures are often very close in value.

9.4. PARTIAL ASSOCIATION MEASURES

We observed for the death penalty example in Section 3.1 that marginal two-way associations can differ strikingly from the partial associations obtained when other variables are controlled. Hence, when we control for other variables, it is important to note how the partial association compares in strength to that in the marginal table.

Suppose that there is a partial table relating the partial association between X and Y at each of l levels of a control variable Z. If the association (as described by some ordinal measure) appears to be similar in the various partial tables, it may be useful to pool the measure values into a single summary measure of partial association. It is natural to take a weighted average of the sample values as a summary measure, using weights $\{w_k\}$ satisfying all $w_k \geq 0$ and $\sum w_k = 1$.

Possible choices for $\{w_k\}$ include the following:

1. $w_k = 1/k$. Here equal weight is given to the sample measure in each table.

2. $w_k = n_{++k}/n$. The kth weight is the proportion of the observations that are in the kth partial table.
3. $w_k = (C_k + D_k)/\sum_{a=1}^{l} (C_a + D_a)$, where C_k and D_k are the numbers of concordant and discordant pairs in the kth partial table. These weights are reasonable if the sample measures are based on the $\{C_k\}$ and $\{D_k\}$.
4. $w_k = (1/\hat{\sigma}_k^2)/(\sum_{a=1}^{l} 1/\hat{\sigma}_a^2)$, where σ_k^2 is the variance of the sample measure in the kth table. This scheme approximates the measure in the class of weighted averages that has the smallest variance. We shall consider this approach in Section 10.4 after giving formulas for σ_k^2 for various ordinal measures.

Davis (1967) used the third set of weights in defining a partial analog of the measure gamma. Let $\hat{\gamma}_k = (C_k - D_k)/(C_k + D_k)$ denote the value of gamma in the kth partial table. Then for his choice of weights

$$\hat{\gamma}_{XY \cdot Z} = \sum_k w_k \hat{\gamma}_k = \frac{\sum_k (C_k - D_k)}{\sum_k (C_k + D_k)} \tag{9.15}$$

Since $\sum_k C_k$ and $\sum_k D_k$ are the total number of concordant and discordant pairs within the partial tables, $\hat{\gamma}_{XY \cdot Z}$ is simply the difference between the proportions of these two types of pairs out of the pairs that are untied on X and Y but tied on Z. Partial gamma shares the advantages and disadvantages of the bivariate gamma measure (9.4). A weighted average of Kendall's tau-b values does not have as simple a structure as (9.15), but it tends to be less dependent in value on the choice of categories of X and of Y. For other approaches to measuring partial association for ordinal variables, see Kendall (1970, p. 121), Quade (1974), and Agresti (1977).

It makes sense to calculate a summary partial measure like $\hat{\gamma}_{XY \cdot Z}$ only when the strength of association is approximately the same in each partial table. Otherwise $\hat{\gamma}_{XY \cdot Z}$ may give a misleading indication of the nature of the association in the partial tables. For example, suppose the association between marital satisfaction and size of family is described by $\hat{\gamma}_1 = 0.50$ for men and $\hat{\gamma}_2 = -0.50$ for women. Then $\hat{\gamma}_{XY \cdot Z}$ would be near zero and would give a false impression of the partial association between marital satisfaction and size of family, controlling for sex. Instead, the separate measures for the partial tables should be reported. The discrepancy between the values and the reasons for it are themselves of interest in such a case.

9.5. CATEGORY CHOICE FOR ORDINAL VARIABLES

In the examples in this book the categorical scale of an ordinal variable is regarded as fixed, typically predetermined by the researcher who originally reported the data. The results of the analyses may depend greatly, however, on the way the categories were selected for the ordinal variables. In this section we shall illustrate this point, first for the ordinal bivariate measures and then for ordinal partial measures.

Bivariate Analyses

The effects of category choice are easily demonstrated by redefining categories in a table. For example, Table 9.4 is a 2×2 condensation of the dumping severity data of Table 2.4. The slight and moderate categories of dumping severity have been combined into the single category "some." Also operation severity is categorized as "low" or "high" by combining the first two and the last two operations. When categories are combined, there are necessarily more ties and fewer concordant and discordant pairs. For Table 9.4 the sample gamma value of 0.269 is more than 50 percent larger than the value of 0.170 obtained for Table 2.4. Tau-*b* is not as greatly affected, changing from 0.111 to 0.135.

It is desirable for a measure to be relatively stable with respect to changes in the categorization if it is to be a reliable index of association. As the numbers of rows and columns are increased, there are fewer tied pairs and measures such as gamma and tau-*b* tend to become closer in value to Kendall's tau, which they both equal for continuous variables. We can judge a measure's stability in terms of how close it tends to be to the limiting value that would be obtained for fully ranked data.

According to this criterion, several authors (e.g., Blalock 1974, Reynolds 1974, Agresti 1976, Hastie and Juritz 1981) have noted that gamma fares poorly because of its tendency to become highly inflated as fewer categories are used. Kendall's tau-*b* tends to be much closer than does gamma to the

Table 9.4. Condensation of Table 2.4 Used to Illustrate Effects of Category Choice

	Dumping Severity		
Operation Severity	None	Some	Total
Low	129	71	200
High	111	106	217
Total	240	177	417

underlying value of Kendall's tau. Generally, it is advantageous to have as many of the pairs untied as possible in order to help determine the true underlying proportions of concordant and discordant pairs. Whenever feasible, therefore, one should use several categories for the measurement of ordinal variables. Because of the dependence of measures of association on the numbers of categories and the marginal distributions, it is dangerous to compare values of measures calculated in tables having different dimensions or highly different marginal distributions.

Another advantage of precise measurement of ordinal variables is that tests based on ordinal measures tend to be more powerful when there are relatively fewer tied pairs of observations. For testing the hypothesis of no association, the sample size needed to attain a fixed power at a fixed significance level tends to decrease as r and c increase. See Agresti (1976) for details.

Partial Association

It is also advantageous to choose several categories for ordinal control variables when one wishes to describe partial association. For example, suppose that for an underlying set of three continuous variables, the value of Kendall's tau between X and Y is identical at each fixed level of the control variable Z. We would want a measure such as $\gamma_{XY \cdot Z}$ formed for describing the partial association in the categorical case to be close in value to the "true" underlying partial association value that would be obtained if the measurement were on the continuous scales. The approximation tends to improve as the number of categories of the control variable increases, since then Z is held more nearly constant.

To illustrate this point, suppose a trivariate normal distribution having $\rho_{XY} = 0.64$ and $\rho_{XZ} = \rho_{YZ} = 0.80$ is categorized into a $2 \times 2 \times l$ table with $\pi_{1++} = \pi_{2++} = \pi_{+1+} = \pi_{+2+} = 0.50$. For the continuous distribution the conditional (X, Y) distribution at a fixed level of Z is bivariate normal with correlation $\rho_{XY \cdot Z} = 0$. Hence the underlying value of the Kendall's tau measure between X and Y at a fixed value of Z is also zero. Table 9.5 contains values of Davis's partial gamma and of the version of partial tau-b having weights $\{w_k = \pi_{++k}\}$ for the $2 \times 2 \times l$ table. It shows that these measures tend to approach zero as the number of categories of the control variable Z increases. However, one needs at least $l = 5$ categories to do this well, and (as in the bivariate case) partial gamma tends to be farther from the true underlying value than does partial tau-b. Note that we would probably fail to detect the absence of partial association if we used only two control categories or if we used partial gamma with, say, no more than four control categories.

Table 9.5. Values of Measures of Partial Association for Various $2 \times 2 \times l$ Tables Having an Underlying Trivariate Normal Distribution

l	Marginal Probabilities for Z	Partial tau-b	Partial gamma
2	(0.1, 0.9)	0.341	0.666
2	(0.5, 0.5)	0.144	0.375
3	(0.3, 0.4, 0.3)	0.075	0.196
4	(0.2, 0.3, 0.3, 0.2)	0.042	0.129
5	(0.2, 0.2, 0.2, 0.2, 0.2)	0.023	0.069
10	(0.1 each)	0.002	0.018
∞		0	0

Note: For the underlying normal distribution $\rho_{XY} = 0.64$, $\rho_{XZ} = \rho_{YZ} = 0.80$. $\pi_{i++} = \pi_{+i+} = 0.50$.

Guidelines Regarding Category Choice

Based on our experience using ordinal measures of association, we feel that the following guidelines apply fairly well. These guidelines should not be regarded as theorems, since the exact behavior of the measures depends on the particular distributional structure. However, we believe that researchers too often ignore the dependence of the results of categorical data analyses on the choice of categories.

1. Ordinal categorical measures of association tend to fall closer to their continuous analogs as r and c increase, and as the proportion of tied pairs decreases.
2. When r or c is small, the absolute value of gamma tends to be considerably larger than its underlying value (Kendall's tau) for continuous variables.
3. Ordinal measures become relatively more efficient at detecting non-null associations as r and c increase, since there are fewer tied pairs and standard errors tend to decrease.
4. Measures of partial association for ordinal categorical data tend to become less biased in describing the partial association for underlying continuous variables as more categories are used for the control variables.
5. Different ordinal measures of association and ordinal models typically yield similar conclusions regarding whether there is an association, when used in significance tests. However, the results of these analyses may depend strongly on how the categories are chosen for

measuring those ordinal variables. The results may also be quite different from those obtained by simply treating the variables as nominal.

The comments in this section have been phrased in terms of measures of association. Several of these remarks also apply to association parameters in ordinal models. Tests based on those models tend to be more powerful at detecting associations for larger table sizes, and parameters describing partial association tend to be more meaningful if ordinal control variables are finely measured.

NOTES

Section 9.2

1. For 2×2 tables, Kendall's tau-b is related to the Pearson X^2 statistic by $\hat{\tau}_b^2 = X^2/n$. The measure X^2/n for 2×2 tables is referred to as *phi-squared*. It is a special case of Cramér's V^2.
2. The $\{r_i^X\}$ and $\{r_j^Y\}$ scores for the marginal distributions are called *ridits*. This terminology is introduced in Section 9.3, where properties of the ridits are formally presented.
3. The canonical correlation is an alternative measure of association. It is the maximum correlation obtained out of all ways of assigning scores to the rows and columns. It also equals the largest eigenvalue of $\mathbf{NN}'$ that is less than 1.0, where $\mathbf{N}$ is the $r \times c$ matrix having elements $n_{ij}/\sqrt{n_{i+}n_{+j}}$. See Kendall and Stuart (1973, pp. 588–598) for details. The resulting scores that produce the canonical correlation need not be monotonic, and the results are invariant to the orderings of rows and columns. For similarities between these scores and the estimated scores for the log-multiplicative model (8.1), see Goodman (1981a) and Haberman (1981).

Section 9.3

4. Notice that the mean ridit for a particular row depends on the other rows in the table, since the ridit scores are calculated from the marginal distribution of the ordinal variable. For instance, we would get different values (and possibly even different orderings) for $\hat{R}_i$ and $\hat{R}_k$ if we calculated their values for the $2 \times c$ table, extracted from the original $r \times c$ table, that contains only rows i and k. The orderings are necessarily the

same if the sample conditional distributions on the ordinal response are stochastically ordered. Also the expected values of the sample ridits depend on the proportions of observations in the various rows. The more observations there are in a particular row (relative to other rows), the more influence that row has on the ridits, since that row is a relatively larger part of the response marginal distribution. The ridits are most meaningful when the proportions of observations in the various rows are the same in the sample as in the population.

5. For $2 \times c$ tables the measure

$$(\hat{R}_2 - \hat{R}_1) + 0.5 = \hat{P}(Y_2 > Y_1) + \left(\frac{1}{2}\right)\hat{P}(Y_2 = Y_1) = \frac{(\hat{\Delta} + 1)}{2}$$

estimates the probability that Y_2 exceeds Y_1 in the underlying continuous distributions. It can also be shown to be algebraically equivalent to the mean ridit score for the $\{p_{j(2)}\}$ distribution when $\{p_{j(1)}\}$ is used as the identified distribution for calculating the ridits.

Section 9.5

6. Quade (1974) suggested an explanation for the tendency of gamma to exceed the value of Kendall's tau for an underlying continuum. Suppose that the proportion $\Pi_c/(\Pi_c + \Pi_d)$ of the tied pairs are regarded as concordant and the proportion $\Pi_d/(\Pi_c + \Pi_d)$ of the tied pairs are regarded as discordant. Then gamma is the simple difference between the overall proportions of concordant and discordant pairs. Ordinarily, one would not expect a pair of observations selected from a subpopulation in which at least one of the variables is restricted in range to exhibit as strong an association as a pair of observations selected from the population of untied pairs. Hence if the proportion of tied pairs that are truly concordant for the underlying continuous distribution is closer to 0.5 than is $\Pi_c/(\Pi_c + \Pi_d)$, then gamma will be farther from 0 than will be the true difference τ in proportions of concordant and discordant pairs.

7. Reynolds (1974, 1977 pp. 107–108) has also illustrated the importance of using several categories for ordinal variables in order to accurately measure multivariate relationships. In a somewhat different context Cochran (1968) discussed the reduction in bias as the number of categories of a covariate is increased.

8. Fine measurement of variables can provide problems in loglinear or logit analysis, due to the sparseness of the data in the cells and the increase in the number of parameters. Nevertheless, researchers should realize that gross collapsing of categories of control variables can lead to serious misrepresentations of partial associations.

COMPLEMENTS

1. Suppose that both classifications in a two-way table have the same categories, listed in the same order. To assess the degree to which observations cluster on the diagonal of the $r \times r$ table, one can compare the probability $\Pi_o = \sum_i \pi_{ii}$ that an observation falls on the diagonal to the corresponding probability $\Pi_e = \sum_i \pi_{i+}\pi_{+i}$ that would be expected if the variables were independent. This measures the degree of agreement in the classifications.

   a. Show that the measure *kappa* (Cohen 1960), defined by

   $$\kappa = \frac{(\Pi_o - \Pi_e)}{(1 - \Pi_e)}$$

   equals one when there is complete agreement and zero when there is independence.

   b. Show that kappa treats the variables as nominal. (A weighted version of kappa, discussed by Fleiss et al. 1969, utilizes distance from the main diagonal and is a more useful measure when the classifications are ordinal. It has the same form, but with $\Pi_o = \sum_i \sum_j w_{ij} \pi_{ij}$ and $\Pi_e = \sum_i \sum_j w_{ij} \pi_{i+} \pi_{+j}$, where the $\{w_{ij}\}$ are weights such as $w_{ij} = 1 - |i - j|/r$.)

2. Let m denote the minimum of r and c in a $r \times c$ table.

   a. Show that the maximum attainable value of $\Pi_c - \Pi_d$ is $(m - 1)/m$. Hence the measure tau-c defined by $\tau_c = m(\Pi_c - \Pi_d)/(m - 1)$ can equal one in absolute value for any table size. See Stuart (1953).

   b. Show that $|\tau_b|$ cannot equal one except when $r = c$.

   c. Similarly, show that $|\rho_b|$ can equal one only when $r = c$, but $|\rho_c|$ can equal one for any table size, where ρ_c is the discrete version of Spearman's rho defined by Stuart (1963) as

   $$\rho_c = 1 - 6m^2 \sum \sum \pi_{ij}(r_i^X - r_j^Y)^2/(m^2 - 1)$$

3. Note from Table 9.2*d* that $|\hat{\gamma}|$ can be large for tables that would be judged to have weak association by other measures. Compute $\hat{\tau}_b$ and d_{YX} for that table, and give the special interpretations that they have for 2×2 tables.

4. Consider the ordinal measure of association C/D. Show that for $2 \times c$ tables it simplifies to α and that for 2×2 tables it simplifies to the odds ratio.

5. For a sample set of counts $(n_1, \ldots, n_c)$ in c ordered categories, the average rank in category j (referred to as the jth *midrank*) is

$$a_j = \frac{\left[\left(\sum_{i=1}^{j-1} n_i\right) + 1\right] + \left[\sum_{i=1}^{j} n_i\right]}{2}$$

Show that the jth midrank is related to the jth ridit r_j by

$$a_j = nr_j + 0.5$$

where $n = \sum_i n_i$. Show that r_j and $a_j/(n+1)$ are essentially the same for large n.

6. Refer to the dumping severity data in Table 2.4. Show that $\hat{\gamma} = 0.170$, $d_{YX} = 0.096$, and $\hat{\tau}_b = 0.111$, and interpret the values.

7. Refer to the political ideology data in Table 5.3. Let D, R, and I denote the Democrat, Republican, and Independent levels of party affiliation, respectively. Show that $\hat{\Delta}_{DR} = 0.436$, $\hat{\Delta}_{DI} = 0.049$, $\hat{\Delta}_{IR} = 0.347$, and $\hat{\alpha}_{DR} = 4.37$, $\hat{\alpha}_{DI} = 1.15$, $\hat{\alpha}_{IR} = 3.27$. Show that the mean ridits are $\hat{R}_D = 0.433$, $\hat{R}_I = 0.486$, $\hat{R}_R = 0.656$. Interpret these values, and compare the three party affiliations on political ideology for this sample.

CHAPTER 10

Inference for Ordinal Measures of Association

Chapter 9 considered summary measures of association that utilize the orderings of the categories of an ordinal variable. The measures were used in a purely descriptive manner. This chapter considers statistical inference for the ordinal measures. The first two sections discuss tests of independence, first for the ordinal-ordinal table and then for the ordinal-nominal table. Like the tests of independence based on ordinal models, these tests focus on narrower alternatives to independence than do the standard chi-squared tests of Section 2.2. Section 10.3 contains expressions for large-sample standard errors for ordinal measures of association. Section 10.4 presents a method for testing equality of values of a measure of association for several populations. Finally, Section 10.5 describes a class of tests of conditional independence based on ordinal measures.

10.1. TESTING FOR ASSOCIATION: ORDINAL-ORDINAL TABLES

When we consider potential departures from independence for ordinal variables, in most applications we first envision monotonic associations. We wish to know whether Y tends to increase or tends to decrease, in some sense, as X increases. If association is described by relative numbers of concordant and discordant pairs, then positive and negative associations are characterized by the cases $\Pi_c > \Pi_d$ and $\Pi_c < \Pi_d$, respectively. Then the null hypothesis of independence can be tested against alternatives such as H_a: $\Pi_c - \Pi_d > 0$, H_a: $\Pi_c - \Pi_d < 0$, or H_a: $\Pi_c - \Pi_d \neq 0$. Each of these alternatives is narrower than the general alternative of dependence used in the chi-squared tests of Section 2.2, since $\Pi_c \neq \Pi_d$ implies dependence but the converse is not true.

Tests Based on $C - D$

The quantity $\Pi_c - \Pi_d$ is negative, zero, or positive in precisely the same instances that γ, τ_b, and Δ_{YX} are. For purposes of testing departures from zero, therefore, it is unnecessary to give separate tests for each measure of association. Instead, a single test can be phrased in terms of $C - D$, the common numerator of the sample values of those measures.

For large random samples the difference $C - D$ is approximately normally distributed. Hence a large-sample test of independence can be based on the test statistic

$$z = \frac{(C - D)}{\sigma(C - D)} \tag{10.1}$$

where $\sigma(C - D)$ is the standard error of $C - D$. Under the null hypothesis an expression can be given for $\sigma(C - D)$ that depends only on the marginal frequencies. Given the marginal totals, Kendall (1970, p. 55) showed that

$$\begin{aligned}\sigma^2(C - D) &= \frac{\{n(n-1)(2n+5) - \sum n_{i+}(n_{i+}-1)(2n_{i+}+5) - \sum n_{+i}(n_{+i}-1)(2n_{+i}+5)\}}{18} \\ &+ \frac{\{\sum n_{i+}(n_{i+}-1)(n_{i+}-2)\}\{\sum n_{+j}(n_{+j}-1)(n_{+j}-2)\}}{9n(n-1)(n-2)} \\ &+ \frac{\{\sum n_{i+}(n_{i+}-1)\}\{\sum n_{+i}(n_{+i}-1)\}}{2n(n-1)}\end{aligned} \tag{10.2}$$

A much simpler expression that is nearly identical to $\sigma^2(C - D)$ for large-sample sizes and can also be used in the z-statistic (10.1) is

$$\tilde{\sigma}^2(C - D) = \frac{[1 - \sum p_{i+}^3][1 - \sum p_{+j}^3]n^3}{9} \tag{10.3}$$

As a rough guideline for use of the normal approximation, we suggest that both C and D should exceed 100.

The standard errors $\sigma(C - D)$ and $\tilde{\sigma}(C - D)$ apply only when the variables are independent. Hence they cannot be used in confidence intervals. Alternative tests can be based on ratios such as

$$\frac{\hat{\tau}_b}{\hat{\sigma}(\hat{\tau}_b)} \quad \text{or} \quad \frac{\hat{\gamma}}{\hat{\sigma}(\hat{\gamma})}$$

where $\hat{\sigma}(\hat{\tau}_b)$ and $\hat{\sigma}(\hat{\gamma})$ are standard errors for $\hat{\tau}_b$ and $\hat{\gamma}$; these will be introduced in Section 10.3. Rejecting the null hypothesis in favor of a two-sided alternative for these tests is equivalent to 0 falling outside the corresponding confidence interval. Simon (1978) showed that all measures having numerator $C - D$ have the same efficacy (and hence the same local power) for testing independence.

Attitude toward Abortion Example

To illustrate this ordinal test of independence, we reconsider the data on attitude toward abortion and schooling in Table 9.1. In Section 9.2 we obtained $C = 246{,}960$, $D = 109{,}063$, $\hat{\gamma} = 0.387$, $\hat{\tau}_b = 0.231$, and $d_{YX} = 0.229$ for this table, indicating a positive association in the sample. Calculation of (10.2) for these data yields $\sigma(C - D) = 14{,}530.2$, and hence

$$z = \frac{(246{,}960 - 109{,}063)}{14{,}530.2} = 9.49$$

There is extremely strong evidence of a positive association between these variables. The same conclusion is obtained using the simpler formula (10.3), for which $\tilde{\sigma}(C - D) = 14{,}525.0$ and $z = 9.49$.

Statistics based on $C - D$ are not the only way one can use ordinal measures of association to test independence. One can base tests on other measures, such as the correlation (9.8). See also Complements 10.2 and 10.3. We have also studied two other single degree of freedom tests in Sections 5.1 and 7.1 that are phrased in terms of association parameters of loglinear and logit models. For $2 \times c$ tables, Bartholomew (1959) formulated an approximate likelihood-ratio statistic for testing independence against the alternative that all local log odds ratios exceed zero. Grove (1980) and Robertson and Wright (1981) formulated similar statistics for the alternative that all global log odds ratios exceed zero (i.e., that row 2 is stochastically higher than row 1).

10.2. TESTING FOR ASSOCIATION: ORDINAL-NOMINAL TABLES

The test discussed in this section is invariant to permutations of rows. Hence it is most naturally applied to tables with nominal row variables and ordinal column variables. It can also be used if the rows are levels of an ordinal variable that we want to compare on an ordinal response variable. In that case this test might be preferred to the one given in the previous section if we expect the levels of the ordinal row variable to be stochastically ordered, but not in a monotonic manner, on the column variable (see Complement 10.6).

Kruskal-Wallis Test

When the column variable is a response variable, independence is often expressed as the property that the conditional distributions on that variable are identical for the r rows. For fully ranked data the Kruskal-Wallis test is the best known rank test for comparing r groups (see Lehmann 1975, p. 204). The test presented here is a discrete adaptation of that test, in which the test statistic for the fully ranked case is replaced by one corrected for ties on the response variable. We express the statistic in terms of the sample mean ridits, since the test is designed to detect differences among the r rows in their true mean ridits. The test detects associations (i.e., the probability of rejecting independence converges to 1 as $n \to \infty$) when at least two of the rows differ in their true mean ridits.

Suppose that the sampling is full multinomial (a single random sample) or independent multinomial within the rows. As in Section 9.3, let $\hat{R}_i$ denote the sample mean ridit for row i, when the sample marginal distribution of the column variable is the identified distribution for forming the ridits. The average of the $\{\hat{R}_i\}$ when weighted by the row marginal counts is 0.50, so their variance is

$$\sum_i (\hat{R}_i - 0.50)^2 \left(\frac{n_{i+}}{n}\right)$$

The Kruskal-Wallis statistic for this setting is directly proportional to this variance, namely

$$W = \frac{12n}{(n+1)T} \sum_i n_{+i}(\hat{R}_i - 0.50)^2 \tag{10.4}$$

where T, called the correction factor for ties, is

$$T = 1 - \frac{\sum_i (n_{+i}^3 - n_{+i})}{(n^3 - n)}$$

For large n, T is approximately equal to $1 - \sum_i p_{+i}^3$. It converges upward toward 1.0 as the number of column categories and the dispersion of the sample among them increases. Under the null hypothesis of independence the Kruskal-Wallis statistic has an asymptotic chi-squared distribution with df $= r - 1$. Mohberg et al. (1978) proposed a related statistic that performs quite well for small samples, and Klotz and Teng (1977) suggested an exact test for that situation.

The Kruskal-Wallis test for comparing r groups on an ordinal response has the same purpose as the chi-squared tests of $H_0: \tau_1 = \cdots = \tau_r = 0$ for the row effects models of Sections 5.2 and 7.2. The test statistics for the loglinear and logit models also have asymptotic null chi-squared distributions with df $= r - 1$. Both the Kruskal-Wallis test and the $G^2(I \mid R)$ tests are analogous to the one-way ANOVA F test for comparing r groups. In the latter case, though, the response variable is continuous and is assumed to be normally distributed with the same variance for each group. The F test statistic for the analysis of variance test also behaves asymptotically like a constant multiple of a chi-squared random variable with df $= r - 1$ (see Complement 10.7). The appropriate descriptive measures for comparing the locations of the groups are the $\{\tau_i\}$ for the loglinear and logit row effects models, the mean ridits for the Kruskal-Wallis test, and the means of the responses for the analysis of variance.

Tonsil Size Example

To illustrate the Kruskal-Wallis test, we reconsider the tonsil size data of Table 9.3. In Section 9.3 we obtained the sample mean ridits $\hat{R}_1 = 0.496$ for noncarriers and $\hat{R}_2 = 0.581$ for carriers. Thus $\sum n_{+i}(\hat{R}_i - 0.50)^2 = 0.493$, the correction for ties $T = 0.866$, and the value of the test statistic is $W = 6.83$. The chi-squared sampling distribution has df $= 1$, so the attained significance level is $P = 0.009$, and there is strong evidence that the tonsil size distribution is different for carriers than for noncarriers.

Pairwise Comparisons

Rarely in practice do we believe that the null hypothesis of independence might be true. Yet the conclusion of the Kruskal-Wallis test simply indicates whether at least two of the r rows differ in their mean ridits. Rejection of H_0 in this test suggests two other questions that are usually of greater practical importance:

1. Which pairs of rows have significant differences in their mean ridits? More generally, to what extent can we make inferences about how the rows are ordered on their mean ridits?
2. How large are the differences among the rows? In other words, how strong is the nominal-ordinal association?

To answer the first of these questions, the Kruskal-Wallis test can be repeated for each pair of rows. This test of the null hypothesis that $r = 2$ conditional distributions are identical is a discrete version of the Wilcoxon (or Mann-Whitney) test. The W statistic then has df = 1 and is the square of a statistic that has an asymptotic normal distribution. For a particular pair of rows i and k, the normal statistic is

$$z = \frac{(\hat{R}_i - 0.50)}{\sigma(\hat{R}_i)} \tag{10.5}$$

where 0.50 is the expected value of $\hat{R}_i$ and $\sigma^2(\hat{R}_i)$ is the variance of $\hat{R}_i$ when H_0 is true, namely

$$\sigma^2(\hat{R}_i) = \frac{n_{k+}(n_{i+} + n_{k+} + 1)}{12n_{i+}(n_{i+} + n_{k+})^2} - \frac{n_{k+} \sum (n_{+j}^3 - n_{+j})}{12n_{i+}(n_{i+} + n_{k+} - 1)(n_{i+} + n_{k+})^3}$$

The statistic $z = (\hat{R}_k - 0.50)/\sigma(\hat{R}_k)$, expressed in terms of the mean ridit for the other row in the $2 \times c$ subtable, is the negative of (10.5). In each application of the test a different $2 \times c$ table is used, and hence the ridits and their means must be recomputed each time. The condition of equal mean ridits for a particular pair of rows is equivalent to the condition that $\Delta = 0$ or $\alpha = 1$ for that $2 \times c$ subtable.

If r is large there are several pairwise comparisons, and a multiple comparison procedure can be used to protect the overall error rate. If we make d comparisons, and if we want the overall type I error rate to be no more than p, for instance, we could use the Bonferroni approach whereby the type I error probability for each comparison is p/d.

For the tonsil size data there are only two rows. Hence $\hat{R}_2 = 0.581$ and $\sigma(\hat{R}_2) = 0.031$ give $z = 2.61$, for which $z^2 = 6.83$ is the value of W previously obtained. The advantage of the z-statistic is that its sign gives the direction of the departure from independence. The z-statistic indicates that the carriers have larger tonsil sizes than noncarriers, as measured by the mean ridit measure.

The second question posed above regarding sizes of differences can be answered by estimating the mean ridit differences, or other ordinal-nominal measures of association such as Δ or α, all of which have simple probability interpretations. We constructed and interpreted point estimates of these measures in Section 9.3. Confidence intervals for the population measures require formulas for standard errors that apply more generally than under the null hypothesis. The next section illustrates the construction of confidence intervals for ordinal measures of association.

10.3. CONFIDENCE INTERVALS FOR MEASURES OF ASSOCIATION

Under suitable sampling assumptions most measures of association have asymptotic normal distributions. In this section we state a result due to Goodman and Kruskal (1972) that gives the general form of the asymptotic distribution for full multinomial sampling. We then apply it to obtain standard errors and large-sample confidence intervals for various measures.

Delta Method

If a statistic is a function of other statistics that are jointly asymptotically normally distributed, then under mild conditions that statistic itself is asymptotically normal. The method of obtaining this asymptotic distribution is called the *delta method.* Descriptions of this method are given by Bishop et al. (1975, Sec. 14.6) and in Appendix C. Now, sample measures of association are functions of cell proportions, which are approximately normally distributed for large n. For these measures the delta method takes the following form:

Let $\zeta = \nu/\delta$ denote an arbitrary measure of association, where the numerator ν and denominator δ are certain functions of the population proportions $\{\pi_{ij}\}$. Let

$$\nu'_{ij} = \frac{\partial \nu}{\partial \pi_{ij}}, \qquad \delta'_{ij} = \frac{\partial \delta}{\partial \pi_{ij}}, \qquad \text{and}$$

$$\phi_{ij} = \delta \nu'_{ij} - \nu \delta'_{ij}, \qquad i = 1, \ldots, r, \qquad j = 1, \ldots, c$$

Suppose that $\hat{\zeta}$ denotes the sample value of ζ for full multinomial sampling. Then $\sqrt{n}(\hat{\zeta} - \zeta)$ has asymptotically (as $n \to \infty$) a normal distribution with mean zero and variance

$$\sigma^2 = \frac{\sum_i \sum_j \pi_{ij} \phi_{ij}^2 - \left(\sum_i \sum_j \pi_{ij} \phi_{ij} \right)^2}{\delta^4} \tag{10.6}$$

Thus the asymptotic variance of a sample measure of association depends on the true cell proportions $\{\pi_{ij}\}$ and on the partial derivatives of the measure taken with respect to the $\{\pi_{ij}\}$. In practice the $\{\pi_{ij}\}$ are unknown, and they are replaced by their sample values in the formula for σ^2, thus giving a ML estimate $\hat{\sigma}^2$ of σ^2. The asymptotic distribution of $\sqrt{n}(\hat{\zeta} - \zeta)/\hat{\sigma}$ is also the standard normal distribution. The term $\hat{\sigma}/\sqrt{n}$ is an approximate standard error for $\hat{\zeta}$, and $\hat{\zeta} \pm z_{p/2}\,\hat{\sigma}/\sqrt{n}$ is an approximate $100(1 - p)$ percent confidence interval for ζ.

Asymptotic Variances of Ordinal Measures

We shall now give the expression for ϕ_{ij} in (10.6) for several measures of association. For a set of sample proportions $\{p_{ij}\}$, we can then calculate $\hat{\phi}_{ij}$ for each cell and hence also $\sum \sum p_{ij} \hat{\phi}_{ij}^2$, $\sum \sum p_{ij} \hat{\phi}_{ij}$, and $\hat{\sigma}^2$. It is also a simple matter to write computer programs for calculating $\hat{\sigma}^2$, and the BMDP-4F package computes standard errors for many measures of association. In some of the following formulas for ϕ_{ij} it is convenient to use the notation

$$\pi_{ij}^{(c)} = \sum_{a<i} \sum_{b<j} \pi_{ab} + \sum_{a>i} \sum_{b>j} \pi_{ab}$$

$$\pi_{ij}^{(d)} = \sum_{a<i} \sum_{b>j} \pi_{ab} + \sum_{a>i} \sum_{b<j} \pi_{ab}$$

The term $\pi_{ij}^{(c)}$ is the sum of probabilities for the cells that are concordant when matched with the cell in row i and column j, and $\pi_{ij}^{(d)}$ is the sum of the probabilities for the cells that are discordant when matched with cell (i, j). Figure 10.1 illustrates these probabilities. Recall also that

$$\Pi_c = 2\sum_{i<k} \sum_{j<l} \pi_{ij}\pi_{kl} \quad \text{and} \quad \Pi_d = 2\sum_{i<k} \sum_{j>l} \pi_{ij}\pi_{kl}$$

Gamma

To apply the delta method to gamma, we express it as $\zeta = \gamma = \nu/\delta$ with $\nu = \Pi_c - \Pi_d$ and $\delta = \Pi_c + \Pi_d$. Then

$$\phi_{ij} = 4(\Pi_d \pi_{ij}^{(c)} - \Pi_c \pi_{ij}^{(d)})$$

and $\sum \sum \pi_{ij}\phi_{ij} = 0$, so that $\sigma^2 = \sum \sum \pi_{ij}\phi_{ij}^2/(\Pi_c + \Pi_d)^4$.

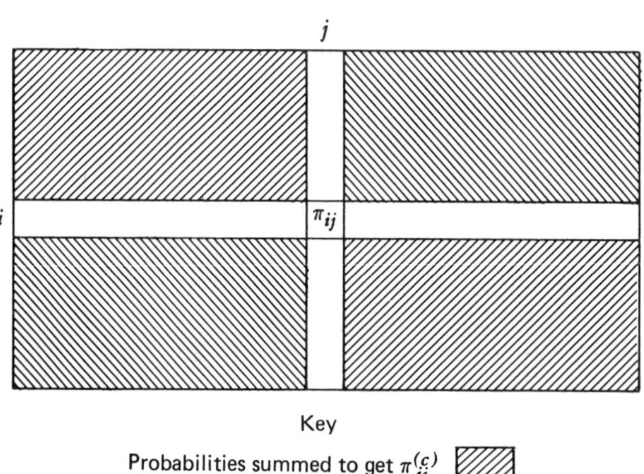

Figure 10.1. Illustration of probabilities summed to get $\pi_{ij}^{(c)}$ and $\pi_{ij}^{(d)}$.

Somers' *d*

$\Delta = v/\delta$ with $v = \Pi_c - \Pi_d$ and $\delta = 1 - \sum_i \pi_{i+}^2$. Here

$$\phi_{ij} = -2\pi_{i+}(\Pi_c - \Pi_d) - 2(1 - \sum_a \pi_{a+}^2)(\pi_{ij}^{(c)} - \pi_{ij}^{(d)})$$

and $\sum \sum \pi_{ij}\phi_{ij} = -2(\Pi_c - \Pi_d)$.

Kendall's tau-*b*

$\tau_b = v/\delta$ with $v = \Pi_c - \Pi_d$ and $\delta = [(1 - \sum_i \pi_{i+}^2)(1 - \sum_j \pi_{+j}^2)]^{1/2}$. Here

$$\phi_{ij} = 2\delta(\pi_{ij}^{(c)} - \pi_{ij}^{(d)}) + v\pi_{+j}\left[\left(1 - \sum_a \pi_{a+}^2\right)\left(1 - \sum_b \pi_{+b}^2\right)\right]^{1/2}$$

$$+ v\pi_{i+}\left[\frac{\left(1 - \sum_b \pi_{+b}^2\right)}{\left(1 - \sum_a \pi_{a+}^2\right)}\right]^{1/2}$$

Alpha

The measure $\alpha = P(Y_2 > Y_1)/P(Y_1 > Y_2)$ for $2 \times c$ tables can be written as $\alpha = v/\delta$ with $v = \sum \sum_{a<b} \pi_{1a}\pi_{2b}$ and $\delta = \sum \sum_{a>b} \pi_{1a}\pi_{2b}$. Here

$$\phi_{1j} = \delta \sum_{b>j} \pi_{2b} - v \sum_{b<j} \pi_{2b} \quad \text{and} \quad \phi_{2j} = \delta \sum_{a<j} \pi_{1a} - v \sum_{a>j} \pi_{1a}$$

$j = 1, \ldots, c$ and $\sum \sum \pi_{ij}\phi_{ij} = 0$. The measure $\log \hat{\alpha}$ converges more rapidly to normality than does $\hat{\alpha}$. The asymptotic variance $\tilde{\sigma}^2$ of $\sqrt{n} \log \hat{\alpha}$ is related to the asymptotic variance σ^2 of $\sqrt{n}\, \hat{\alpha}$ by $\tilde{\sigma}^2 = \sigma^2/\alpha^2$. This can be used to get a confidence interval for $\log \alpha$, and the end points of that interval can be exponentiated to get a confidence interval for α.

Delta

The measure $\Delta = P(Y_2 > Y_1) - P(Y_1 > Y_2)$ for $2 \times c$ tables has $v = \sum\sum_{a<b} \pi_{1a}\pi_{2b} - \sum\sum_{a>b} \pi_{1a}\pi_{2b}$ and $\delta = \pi_{1+}\pi_{2+}$. Here

$$\phi_{1j} = \delta\left(\sum_{b>j} \pi_{2b} - \sum_{b<j} \pi_{2b}\right) - v\pi_{2+} \qquad \text{and}$$

$$\phi_{2j} = \delta\left(\sum_{a<j} \pi_{1a} - \sum_{a>j} \pi_{1a}\right) - v\pi_{1+}, \qquad j = 1, \ldots, c$$

and $\sum \sum \pi_{ij}\phi_{ij} = 0$.

Other Measures

Goodman and Kruskal (1963, 1972) gave the asymptotic variances of several measures of association for independent multinomial sampling as well as for full multinomial sampling. See Note 10.4 for the appropriate formula for the case of proportional sampling within the rows. An alternative approach for obtaining the asymptotic variances using compounded functions of matrices is described in Forthofer and Koch (1973). Semenya and Koch (1980) used this approach to obtain the asymptotic covariance matrix of the mean ridits. Their matrix can be used to obtain the variance of the difference $\hat{R}_i - \hat{R}_k$ between two mean ridits, which is the same as the variance of $\hat{R}_i - \hat{R}_k + 0.5$, an estimate of $P(Y_i > Y_k)$ in the underlying continuous population. This probability is also estimated by $(\hat{\Delta}_{ik} + 1)/2$, where $\hat{\Delta}_{ik}$ is the value of $\hat{\Delta}$ computed for the $2 \times c$ table consisting of rows i and k. The standard error of this measure is half that of $\hat{\Delta}$ given previously. The asymptotic covariance matrix of the mean ridits can also be obtained using a generalized version of the delta method (see Complement 10.8).

Attitude on Abortion Example

We illustrate the calculation of an asymptotic standard error for the measure gamma applied to the attitude on abortion data in Table 9.1. We estimate π_{ij} by $p_{ij} = n_{ij}/n$, Π_c by $P_c = 2C/n^2$, and Π_d by $P_d = 2D/n^2$. Here $C = 246{,}960$, $D = 109{,}063$, and $n = 1425$, so that $P_c = 0.243$ and $P_d = 0.107$. Then

$$\hat{\sigma}^2 = \frac{\sum \sum p_{ij} \hat{\phi}_{ij}^2}{(P_c + P_d)^4}$$

where $\hat{\phi}_{ij} = 4(P_d \hat{\pi}_{ij}^{(c)} - P_c \hat{\pi}_{ij}^{(d)})$ can be expressed as

$$\hat{\phi}_{ij} = \frac{8[D(n\hat{\pi}_{ij}^{(c)}) - C(n\hat{\pi}_{ij}^{(d)})]}{n^3}$$

The term $n\hat{\pi}_{ij}^{(c)}$ represents the *number* of observations that are concordant when matched with cell (i, j). For instance, $n\hat{\pi}_{11}^{(c)} = 126 + 426 + 21 + 138 = 711$. Table 10.1 lists n_{ij}, $n\hat{\pi}_{ij}^{(c)}$, $n\hat{\pi}_{ij}^{(d)}$, and $\hat{\phi}_{ij}$ for the attitude on abortion data. Substitution of these into the estimated variance formula gives $\hat{\sigma}^2 = 1.910$. Hence the standard error of $\hat{\gamma} = 0.387$ is $\hat{\sigma}/\sqrt{n} = 0.037$, and an approximate 95 percent confidence interval for γ is $\hat{\gamma} \pm 1.96\hat{\sigma}/\sqrt{n}$, or (0.315, 0.459). We conclude that there is a positive association of moderate size between amount of schooling and attitude on abortion.

Table 10.1. Calculations for Terms in Standard Error of Gamma

		Attitude toward Abortion		
Schooling	Term	Disapprove	Middle	Approve
Less than high school				
	n_{1j}	209	101	237
	$n\hat{\pi}_{1j}^{(c)}$	711	564	0
	$n\hat{\pi}_{1j}^{(d)}$	0	167	314
	$\hat{\phi}_{1j}$	0.214	0.056	−0.214
High school				
	n_{2j}	151	126	426
	$n\hat{\pi}_{2j}^{(c)}$	159	347	310
	$n\hat{\pi}_{2j}^{(d)}$	338	253	37
	$\hat{\phi}_{2j}$	−0.183	−0.068	0.068
More than high school				
	n_{3j}	16	21	138
	$n\hat{\pi}_{3j}^{(c)}$	0	360	587
	$n\hat{\pi}_{3j}^{(d)}$	890	663	0
	$\hat{\phi}_{3j}$	−0.608	−0.344	0.177

10.4. COMPARING AND POOLING MEASURES

Inferences can also be made about partial associations using the results of the past three sections. Suppose that we are interested in the association between X and Y at l levels of a control variable Z. Two questions regarding the partial associations are often important.

1. Can we assume that the strength of association is the same at each of the l levels of Z?
2. If we can assume common association, is it nonnull?

For nominal variables the first question is addressed by the goodness-of-fit test of the loglinear model (XY, XZ, YZ), whereas the second is addressed using the statistic $G^2[(XZ, YZ) \mid (XY, XZ, YZ)]$. Analogous tests based on models for ordinal variables were considered in Sections 5.3 and 7.3. This chapter considers tests based on ordinal measures of association. This section describes a test for comparing the l partial associations. Section 10.5

discusses methods for testing the hypothesis of conditional independence between X and Y, controlling for Z.

Let ζ_k, $k = 1, \ldots, l$ denote the values of some measure of association ζ between X and Y that we would like to compare among the l levels of Z. For instance, if X and Y are ordinal, ζ_k could be the value of gamma or tau-b at the kth level of Z. Suppose that the sampling scheme is some form of multinomial. We shall treat the sample sizes $\{n_k = n_{++k}\}$ from the various levels of Z as fixed. Thus samples from different layers of Z can be regarded as independent, since even if the original sampling was full multinomial, conditioning on the $\{n_k\}$ gives independent multinomial layer samples. Let σ_k^2 denote the asymptotic variance (as $n_k \to \infty$) of $\sqrt{n_k}\hat{\zeta}_k$, where $\hat{\zeta}_k$ denotes the sample value of ζ_k in level k of Z.

Comparing Two Values

An estimate of the difference between a particular pair ζ_h and ζ_k is given by the sample difference $\hat{\zeta}_h - \hat{\zeta}_k$. When n_h and n_k are large, this difference has approximately a normal distribution with standard error $\sqrt{\hat{\sigma}_h^2/n_h + \hat{\sigma}_k^2/n_k}$, where $\hat{\sigma}_h^2$ and $\hat{\sigma}_k^2$ are the sample versions of (10.6) applied to the measure of interest. An approximate large-samples confidence interval for $\zeta_h - \zeta_k$ is

$$(\hat{\zeta}_h - \hat{\zeta}_k) \pm z_{p/2}\sqrt{\hat{\sigma}_h^2/n_h + \hat{\sigma}_k^2/n_k}$$

Similarly, a test of H_0: $\zeta_h = \zeta_k$ can be based on the test statistic

$$z = \frac{(\hat{\zeta}_h - \hat{\zeta}_k)}{\sqrt{\hat{\sigma}_h^2/n_h + \hat{\sigma}_k^2/n_k}}$$

This statistic has approximately a standard normal null distribution for large n_h and n_k.

Comparing Several Values

Now consider the more general hypothesis H_0: $\zeta_1 = \cdots = \zeta_l$ of equal partial associations, for $l \geq 2$. Assuming H_0 is true, an estimate of the common partial association is given by a weighted average $\sum w_k \hat{\zeta}_k$ of the $\{\hat{\zeta}_k\}$, where all $w_k \geq 0$ and $\sum w_k = 1$. Various choices for the weights were suggested in Section 9.4. Here denote by $\hat{\zeta}_{XY\cdot Z}$ the weighted average with

$$w_k = \frac{(n_k/\hat{\sigma}_k^2)}{\sum_{a=1}^{l} (n_a/\hat{\sigma}_a^2)}$$

The weight assigned to a particular $\hat{\zeta}_k$ is inversely proportional to its estimated variance. The measure $\hat{\zeta}_{XY \cdot Z}$ approximates the weighted average of the $\{\hat{\zeta}_k\}$ that has the smallest variance.

Now consider the statistic

$$V = \sum \frac{n_k(\hat{\zeta}_k - \hat{\zeta}_{XY \cdot Z})^2}{\hat{\sigma}_k^2} \tag{10.7}$$

where the squared difference between each sample estimate $\hat{\zeta}_k$ and the pooled value $\hat{\zeta}_{XY \cdot Z}$ is divided by the estimated variance of $\hat{\zeta}_k$. For large $\{n_k\}$, this statistic has approximately a chi-squared null distribution with $\text{df} = l - 1$. Large values for V give evidence that at least two of the measures are unequal. If the null hypothesis of equal partial associations is rejected, pairwise comparisons can be made using z-tests or confidence intervals for differences between values.

If it is reasonable to assume $\zeta_1 = \cdots = \zeta_l$, the measure $\hat{\zeta}_{XY \cdot Z}$ is valuable as a summary index of partial association. It gives a pooling of the information from the l levels into a single estimate of the common partial association. Under the sampling assumptions we have made, it is asymptotically normally distributed with estimated asymptotic variance

$$\hat{\sigma}^2(\hat{\zeta}_{XY \cdot Z}) = \left(\sum \frac{n_k}{\hat{\sigma}_k^2}\right)^{-1} \tag{10.8}$$

Hence an approximate confidence interval for the true partial association $\zeta_{XY \cdot Z}$ is

$$\hat{\zeta}_{XY \cdot Z} \pm \frac{z_{p/2}}{\sqrt{\sum n_k / \hat{\sigma}_k^2}}.$$

Dumping Severity Example

The discussion in this section has been very general, in the sense that ζ could be a measure of association for ordinal-ordinal tables, ordinal-nominal tables, or even nominal-nominal tables. Also the control variable Z could be of any scale, since the partial association measure $\zeta_{XY \cdot Z}$ does not depend on the ordering of the control categories. We now illustrate these results for the ordinal measure gamma applied to Table 4.9, the three-dimensional table in which the association between operation and dumping severity is given for four hospitals.

The sample values of gamma for the four hospitals are $\hat{\gamma}_1 = 0.135$, $\hat{\gamma}_2 = 0.385$, $\hat{\gamma}_3 = 0.068$, and $\hat{\gamma}_4 = 0.084$. Using (10.6), we obtain the standard error values of 0.110, 0.120, 0.149, and 0.144. Hence the weights are $w_1 = 0.335$, $w_2 = 0.284$, $w_3 = 0.184$, and $w_4 = 0.197$, and the estimate of an assumed common gamma value for each hospital is $\hat{\gamma}_{XY \cdot Z} = \sum w_k \hat{\gamma}_k = 0.184$. Its standard error is $\hat{\sigma}(\hat{\gamma}_{XY \cdot Z}) = 0.064$, and an approximate 95 percent confidence interval for the assumed common value $\gamma_{XY \cdot Z}$ is (0.059, 0.309).

Before using $\hat{\gamma}_{XY \cdot Z}$ as a summary of the measures for the partial tables, though, we should first check that H_0: $\gamma_1 = \gamma_2 = \gamma_3 = \gamma_4$ is tenable. The chi-squared statistic for testing this hypothesis is $V = \sum n_k(\hat{\gamma}_k - \hat{\gamma}_{XY \cdot Z})^2/\hat{\sigma}_k^2 = 4.10$, based on df $= 4 - 1 = 3$. Although there is some evidence of a stronger association for the second hospital, the data do permit the assumption of a common partial association.

Similar results are obtained using other ordinal measures. For example, the values of $\hat{\tau}_b$ (with standard errors in parentheses) are $\hat{\tau}_{b1} = 0.086$ (0.071), $\hat{\tau}_{b2} = 0.253$ (0.081), $\hat{\tau}_{b3} = 0.046$ (0.101), and $\hat{\tau}_{b4} = 0.054$ (0.093). From these we obtain weights $w_1 = 0.355$, $w_2 = 0.268$, $w_3 = 0.172$, $w_4 = 0.206$, the summary partial measure $\hat{\tau}_{XY \cdot Z} = 0.118$ (with standard error 0.042), and $V = 3.93$ for testing H_0: $\tau_{b1} = \tau_{b2} = \tau_{b3} = \tau_{b4}$.

10.5. TESTING FOR PARTIAL ASSOCIATION

Partial associations can be drastically different from marginal associations. Even if we have observed a marginal association between X and Y, it is important to study whether the variables are conditionally independent when other variables are controlled.

Assumed Common Association

In the previous section we gave a test of H_0: $\zeta_1 = \cdots = \zeta_l$ and an estimate of the common value $\zeta_{XY \cdot Z}$. If the assumption of a common association seems reasonable, then an obvious way to test the hypothesis of conditional independence is through the statistic

$$z = \frac{\hat{\zeta}_{XY \cdot Z}}{\hat{\sigma}(\hat{\zeta}_{XY \cdot Z})} \tag{10.9}$$

This test statistic has an asymptotic standard normal distribution if $\zeta_{XY \cdot Z} = 0$. It gives a consistent test (i.e., the probability it rejects H_0 converges to 1.0 as $n \rightarrow \infty$) for the alternative hypothesis H_a: $\zeta_{XY \cdot Z} \neq 0$. For the dumping severity data classified by hospital (Table 4.9), we obtained $\hat{\gamma}_{XY \cdot Z} = 0.184$ and $\hat{\sigma}(\hat{\gamma}_{XY \cdot Z}) = 0.064$ in the previous section, so that $z = 0.184/0.064 = 2.88$ gives strong evidence of a positive partial association.

An alternative statistic that is simpler because it is based on a null variance is as follows: For each partial table compute $C_k - D_k$, and compute its variance $\sigma^2(C_k - D_k)$ under the assumption that X and Y are independent at that level of Z [using (10.2) or (10.3)]. Then the hypothesis that X and Y are conditionally independent, controlling for Z, can be tested using

$$z = \frac{\sum (C_k - D_k)}{\sqrt{\sum \sigma^2(C_k - D_k)}} \tag{10.10}$$

This statistic is analogous to one presented by Mantel and Haenszel (1959) (see Note 10.8) for $2 \times 2 \times l$ tables. It equals 2.58 when applied, using (10.3), to the dumping severity data.

No Assumed Common Association

Notice that the approach just discussed gave strong evidence of a positive partial association between dumping severity and operation, even though the evidence was not especially strong in all partial tables. Only the second hospital gave significant evidence ($z = \sqrt{n_2}\hat{\gamma}_2/\hat{\sigma}_2 = 3.2$) of an association. Since the four associations all had the same sign, however, their cumulative effect when pooled into the partial measure $\hat{\gamma}_{XY \cdot Z}$ was to suggest strongly a positive partial association. This approach is not sensitive to departures from conditional independence in which the association varies in sign across the levels of Z. For example, if $\hat{\zeta}_1 = 0.75$ and $\hat{\zeta}_2 = -0.68$ when $l = 2$, then $\hat{\zeta}_{XY \cdot Z}$ could be close to zero. Thus z-statistics such as (10.9) might not give evidence of a partial association, even though there is strong evidence within each partial table. This is because (10.9) is consistent not for the alternative hypothesis, H_a: at least one $\zeta_k \neq 0$, but for the narrower alternative that $\zeta_{XY \cdot Z} \neq 0$. Hence tests such as (10.9) and (10.10) should be used only when it can be assumed that the association is constant in the partial tables.

Alternative tests can be formulated that do not require the assumption of constant association in the partial tables. For instance, let $z_k = \sqrt{n_k}(\hat{\zeta}_k/\hat{\sigma}_k)$, $k = 1, \ldots, l$, and consider the statistic

$$\sum z_k^2 = \sum \frac{n_k \hat{\zeta}_k^2}{\hat{\sigma}_k^2} \tag{10.11}$$

Again regarding the samples in the partial tables as independent, when $\zeta_1 = \cdots = \zeta_l = 0$ we can treat $\sum z_k^2$ as an asymptotic chi-squared statistic having df $= l$.

The statistic (10.11) tests whether $\zeta_1 = \cdots = \zeta_l = 0$, whereas the statistic (10.9) tested whether $\zeta_{XY \cdot Z} = 0$. Note that $\zeta_{XY \cdot Z} = 0$ is equivalent to $\zeta_1 = \cdots = \zeta_l = 0$ only under the added assumption that $\zeta_1 = \cdots = \zeta_l$, which was made to use statistic (10.9). Hence we observed that the test based on $\hat{\zeta}_{XY \cdot Z}$ is at a disadvantage if that assumption is very false. The test based on $\sum z_k^2$ has its own disadvantage, however. Since it is insensitive to the signs of the sample measures, it cannot take advantage of similar results from the partial tables like the test based on $\hat{\zeta}_{XY \cdot Z}$ can. The test based on $\hat{\zeta}_{XY \cdot Z}$ focuses the partial association on a single degree of freedom, and it also allows us to test against the one-sided alternative hypotheses $\mathrm{H_a}$: $\zeta_{XY \cdot Z} < 0$ and $\mathrm{H_a}$: $\zeta_{XY \cdot Z} > 0$. If the assumption of a common association is reasonable, then the statistic (10.9) is preferable.

The distinction between statistics (10.9) and (10.11) is the same as the distinction between the statistics $G^2[(XZ, YZ) \mid (XZ, YZ, XY)]$ and $G^2[(XZ, YZ)]$ for standard loglinear models. In the first case the conditional independence of X and Y is tested under the assumption of no three-factor interaction, whereas in the second case no such assumption is made. In fact an analogy can be made here to the decomposition

$$G^2[(XZ, YZ)] = G^2[(XZ, YZ) \mid (XZ, YZ, XY)] + G^2[(XZ, YZ, XY)]$$

that is used in testing conditional independence for nominal variables. It can be easily seen that

$$\sum z_k^2 = \left(\frac{\hat{\zeta}_{XY \cdot Z}}{\hat{\sigma}(\hat{\zeta}_{XY \cdot Z})} \right)^2 + V, \tag{10.12}$$

where $z_k^2 = n_k(\hat{\zeta}_k / \hat{\sigma}_k)^2$. Here $\sum z_k^2$ is the chi-squared statistic having $\mathrm{df} = l$ for testing conditional independence. The statistic $(\hat{\zeta}_{XY \cdot Z}/\hat{\sigma}(\hat{\zeta}_{XY \cdot Z}))^2$ is the square of statistic (10.9), and it gives a chi-squared test with $\mathrm{df} = 1$ for testing conditional independence under the assumption that the partial associations are constant. The statistic V is the chi-squared statistic with $\mathrm{df} = l - 1$ for testing no three-factor interaction as defined by equality of the partial associations $\zeta_1, \ldots, \zeta_l$. For gamma applied to the dumping severity data, for instance, $\sum z_k^2 = 12.4$ based on $\mathrm{df} = 4$, $V = 4.1$ based on $\mathrm{df} = 3$, and $(\hat{\gamma}_{XY \cdot Z}/\hat{\sigma}(\hat{\gamma}_{XY \cdot Z}))^2 = (2.88)^2 = 8.3$ based on $\mathrm{df} = 1$.

The approach of summing chi-squared statistics from partial tables [as in (10.11)] can be used with partial tables other than ordinal-ordinal ones. For example, Section 10.2 discussed the Kruskal-Wallis test for association in an $r \times c$ nominal-ordinal table. The test statistic (10.4) has asymptotically a chi-squared distribution under $\mathrm{H_0}$ with $\mathrm{df} = r - 1$. If there are independent nominal-ordinal partial tables at each of l levels of a control variable, the sum of these chi-squared statistics has an asymptotic chi-squared distribution with $\mathrm{df} = l(r - 1)$ under the hypothesis that X and Y are conditionally independent.

Models for Measures of Association

The measures of association discussed in Chapters 9 and 10 are not parameters in structural models. Hence the analyses in these chapters have had more of a nonparametric flavor than those discussed in previous chapters. It is sometimes useful, however, to phrase models in terms of these measures of association.

In Section 10.4 the estimation of an assumed constant partial association is equivalent to fitting the simple model

$$\zeta_i = \alpha, \qquad i = 1, \ldots, l$$

where we denoted $\alpha = \zeta_{XY \cdot Z}$. Fitting this model by the WLS method described in Appendix A produces a residual chi-squared statistic based on $\text{df} = l - 1$. The WLS estimate of α is simply the value of $\hat{\zeta}_{XY \cdot Z}$ corresponding to the weighted average of $\{\hat{\zeta}_i\}$ having weights inversely proportional to sampling variances. The residual goodness-of-fit statistic is the statistic V in (10.7). Forthofer and Koch (1973) provided details of this approach. More general models, such as $\zeta_i = \alpha + \beta z_i$ for fixed scores $\{z_i\}$ on Z, can be easily fitted using the WLS method. Forthofer and Lehnen (1981) gave good discussions of a wide variety of models that can be fitted using WLS.

Other models have been proposed in which ordinal measures of association occur as parameters. For instance, recall that Kendall's tau-*b* is a correlation coefficient computed for sign scores of differences between X values and differences between Y values for pairs of observations. Hawkes (1971) and Ploch (1974) suggested the multivariate analysis of ordinal variables using linear models for sign scores. They modeled the sign scores for the difference between a pair of observations on the response as a linear function of corresponding sign scores for explanatory variables. Partial regression coefficients and partial correlation coefficients for this model are multivariate analogs of Somers' *d* and Kendall's tau-*b*.

In many applications it is reasonable to expect the difference between the conditional distributions on the response to increase in some sense when we increase the distance between two levels of an explanatory variable. For the case of an ordinal response variable, Schollenberger et al. (1979) formulated a model in which the logit for the probability of concordance for a pair of observations is linearly related to the distance between the observations on the explanatory variables.

The pair-score models can be fitted using WLS. Like the mean response models discussed in Section 8.3, they are formulated in terms of summary responses and do not provide a structure in which conditions such as independence are embedded.

NOTES

Section 10.1

1. To get expression (10.3) from (10.2), the third term in (10.2) is dropped, since it is of smaller order of magnitude than the first two terms, and the -1, -2, and $+5$ terms are omitted from the first two terms. The resulting expression is asymptotically equivalent to (10.2), in the sense that $\sigma^2(C-D)/\tilde{\sigma}^2(C-D) \to 1$ as $n \to \infty$.

Section 10.2

2. Koch et al. (1981) suggested alternative chi-squared statistics for testing independence in ordinal-nominal tables. These require assigning scores to the levels of the ordinal variable.

Section 10.3

3. In some situations the standard error formulas must be used with caution. For instance, suppose that $|\gamma| = 1$. Then $|\hat{\gamma}| = 1$ with probability one and $\sigma = 0$. However, $\hat{\sigma} = 0$ occurs whenever $|\hat{\gamma}| = 1$, which is not an unusual sample result for small sample sizes even if $|\gamma| < 1$. Gans and Robertson (1981a, 1981b) showed that gamma has a tendency also to converge slowly to normality and to have general distributional irregularity, bias, and skewness problems.

4. Suppose that there is independent multinomial sampling within the rows with proportional sampling; that is, the proportion sampled from the ith row equals the proportion π_{i+} of the population classified in row i. In that case the expression for σ^2 in (10.6) is replaced by

$$\sigma^2 = \frac{\sum_i \sum_j \pi_{ij}\phi_{ij}^2 - \sum_i \pi_{i+}\left(\sum_j \pi_{j(i)}\phi_{ij}\right)^2}{\delta^4}$$

where $\pi_{j(i)} = \pi_{ij}/\pi_{i+}$. For the delta and alpha measures, $\sum \pi_{j(1)}\phi_{1j} = \sum \pi_{j(2)}\phi_{2j} = 0$, and the variance is the same as it is for full multinomial sampling.

5. Fleiss (1982) showed that if a measure has exactly the same form when expressed in terms of $\{p_{ij}\}$ as it does when expressed in terms of $\{n_{ij}\}$ (i.e., substituting n_{ij} for p_{ij} in its formula gives exactly the same value), then $\sum \sum \pi_{ij}\phi_{ij} = 0$ in (10.6). This property holds for the measures gamma, delta, alpha, and the odds ratio.

Section 10.4

6. The convergence of $\hat{\zeta}_h - \hat{\zeta}_k$ to normality and the convergence of V to a chi-squared distribution with df $= l - 1$ is based on the sample sizes $\{n_a \rightarrow \infty\}$ such that $n_a/\sum n_k \rightarrow \lambda_a$ with $0 < \lambda_a < 1$ for $a = 1, \ldots, l$.

Section 10.5

7. Mantel (1963) described tests having the form (10.9) in which the component association measures $\{\zeta_k\}$ are correlation coefficients.
8. Refer to statistic (10.10). Suppose that each partial table has size 2×2, and let $\hat{m}_{11k} = n_{1+k}n_{+1k}/n_{++k}$ denote the estimated expected value assuming conditional independence between X and Y, $k = 1, \ldots, l$. Another way of pooling the evidence about association, assuming the same association in each table, is through the statistic

$$z = \frac{\sum_k (n_{11k} - \hat{m}_{11k})}{\sqrt{\sum_k \hat{\sigma}^2(n_{11k})}}$$

where $\hat{\sigma}^2(n_{11k}) = n_{1+k}n_{2+k}n_{+1k}n_{+2k}/[n_{++k}^2(n_{++k} - 1)]$. The square of this z-statistic, with an added continuity correction, was proposed by Mantel and Haenszel (1959) for testing independence between X and Y, controlling for Z. Under H_0 this squared statistic also has an asymptotic chi-squared distribution with df $= 1$.

COMPLEMENTS

1. Derive the expressions given for ϕ_{ij} that are used in the asymptotic variance formula for:
   a. gamma
   b. Somers' d
   c. tau-b
   d. delta
2. An $r \times 2$ table is obtained when we compare r groups on a dichotomous response.
   a. Show that the Pearson X^2 statistic for testing independence in a $r \times 2$ table can be expressed as

$$X^2(I) = \left(\frac{1}{p_{+1}p_{+2}}\right) \sum n_{i+}(p_{1(i)} - p_{+1})^2$$

b. Suppose that the r rows are ordered and that scores $\{x_i\}$ are assigned to the rows. For example, we may record the proportion $p_{1(i)}$ of positive responses to a drug administered at dosage level x_i. Consider the linear model $\pi_{1(i)} = \alpha + \beta x_i$, which we denote by (L). Note that the independence model (I) is the special case of (L) with $\beta = 0$. Let $\hat{\pi}_{1(i)}$ be the predicted value of $\pi_{1(i)}$ at x_i for (L). Show that the ordinary least squares fit gives the prediction equation

$$\hat{\pi}_{1(i)} = p_{+1} + b(x_i - \bar{x})$$

where

$$\bar{x} = \sum \frac{n_{i+} x_i}{n} \quad \text{and}$$

$$b = \frac{\sum n_{i+}(p_{1(i)} - p_{+1})(x_i - \bar{x})}{\sum n_{i+}(x_i - \bar{x})^2}$$

c. Show that the goodness of fit of the linear model can be tested using

$$X^2(L) = \left(\frac{1}{p_{+1}p_{+2}}\right) \sum n_{i+}(p_{1(i)} - \hat{\pi}_{1(i)})^2$$

which has an asymptotic chi-squared distribution with df $= r - 2$ if the $\{\pi_{1(i)}\}$ are actually related to $\{x_i\}$ as given in the linear model.

d. Given that the linear model holds, a test of independence (H_0: $\beta = 0$) that utilizes the ordering of the rows can be based on

$$X^2(I \mid L) = X^2(I) - X^2(L)$$

which has an asymptotic chi-squared distribution with df $= 1$ when H_0 is true. Show that

$$X^2(I \mid L) = \left(\frac{b^2}{p_{+1}p_{+2}}\right) \sum n_{i+}(x_i - \bar{x})^2$$

See Cochran (1954) and Armitage (1955) for details.

3. In an $r \times c$ table with ordered rows and ordered columns, we could assign ordered scores to the rows and to the columns, and hypothesize a linear trend $E(Y \mid X = x) = \alpha + \beta x$. If b is the least squares estimator of β, show that the statistic $X^2(I)$ for testing independence can be

partitioned into

$$X^2(I) = X^2(L) + X^2(I \mid L)$$

where $X^2(I \mid L) = b^2/\text{Var}(b)$. Here $X^2(L)$ has df $= rc - r - c$ for testing the goodness of fit of the linear model, and $X^2(I \mid L)$ has df $= 1$ for testing the hypothesis of independence, given that the linear model holds. The test based on $X^2(I \mid L)$ (or on the z-statistic given by its square root) is an alternative way of testing for association to the z-tests given in this chapter based on measures of association or on $C - D$. See Yates (1948) and Everitt (1977, pp. 51–55). Also see Section 8.3 for a weighted least squares approach with this model.

4. The data relating dumping severity to operation were analyzed using loglinear and logit uniform association models in Sections 5.1 and 7.1.

a. Show that the test of independence using (10.1) yields $z = 2.55$ for those data. Compare this result to those obtained in conditional tests of independence for the models.

b. Note the difference between the result in (a) and the result of the standard chi-squared test (using $G^2(I)$) that treats the variables as nominal.

c. Show that for gamma, $\hat{\sigma}^2 = 1.75$, and an approximate 95 percent confidence interval is (0.043, 0.297).

d. Discuss how the methods based on ordinal measures of association have lead to similar substantive conclusions for these data as were obtained using the models.

5. The data relating political ideology and party affiliation were analyzed using loglinear and logit row effects models in Sections 5.2 and 7.2.

a. Show that the test of independence using the Kruskal-Wallis statistic yields $W = 96.6$ based on df $= 2$. Compare this result to the corresponding tests of H_0: $\tau_1 = \tau_2 = \tau_3 = 0$ having df $= 2$ for the models.

b. Show that when Democrats and Independents alone are compared using the Wilcoxon statistic (10.5), the sample mean ridits are $\hat{R}_1 = 0.470$ and $\hat{R}_2 = 0.525$, and $\sigma(\hat{R}_1) = 0.010$ and $z = -3.00$. Conclude from this and the other comparisons that the groups all differ on political ideology, with Democrats tending to be the least conservative and Republicans tending to be the most conservative.

c. Discuss how methods based on ridits have lead to similar substantive conclusions for these data as were obtained using the models.

6. Suppose that for the row variable "drug dosage (low, medium, high)" and the column variable "effect on patient's condition (negative, none, positive)," the conditional distributions on the column variable are expected to exhibit a pattern such as the following:

	−	None	+
L	0.30	0.40	0.30
M	0.10	0.30	0.60
H	0.30	0.40	0.30

Indicate why it is preferable to test independence using statistic (10.4) rather than (10.1), even though both variables are ordinal.

7. Recall that the F test statistic in the one-way analysis of variance has, under the null hypothesis of r equal means, an F distribution with degrees of freedom $r - 1$ and $n - r$. Show that this distribution is asymptotically (as the total sample size $n \to \infty$) that of a chi-squared random variable with $\text{df} = r - 1$ divided by its degrees of freedom. Note the similarity to the asymptotic distributions obtained using the Kruskal-Wallis test or using the conditional tests of independence with the loglinear or logit row effects models to compare r groups on ordinal response distributions.

8. Suppose that each element of a vector is some function of another vector that is asymptotically multivariate normal. There is a multivariate version of the delta method formula (C.1) in Appendix C that gives the asymptotic covariance matrix of the transformed vector (e.g., see Rao 1973, p. 388, and Bishop et al. 1975, p. 493). Using this result the vector $\hat{R}' = (\hat{R}_1, \ldots, \hat{R}_{r-1})$ of sample mean ridits in a $r \times c$ table has an asymptotic multivariate normal distribution, with the asymptotic covariance matrix of $\sqrt{n}(\hat{\mathbf{R}} - \mathbf{R})$ equal to $\mathbf{D}'\mathbf{\Sigma}\mathbf{D}$. Here $\mathbf{\Sigma}/n$ is the covariance matrix for the sample proportions (given in Appendix C.2 for the case of full multinomial sampling) and $\mathbf{D}'$ is a $(r - 1) \times rc$ matrix whose kth row contains derivatives of R_k taken with respect to the population proportions. Show that for this matrix $\mathbf{D}$,

$$\begin{aligned} \partial R_k/\partial \pi_{ij} &= 1 - r_{j(k)}, \qquad i \neq k \\ &= 1 - r_{j(i)} + [r_j - R_i]/\pi_{i+}, \qquad i = k \end{aligned}$$

where $r_{j(k)} = \sum_{l<j} \pi_{l(k)} + (\frac{1}{2})\pi_{j(k)}$. Notice that $\hat{\mathbf{R}}$ contains only $r - 1$ of the r sample mean ridits because of the linear constraint $\Sigma\, p_{i+}\hat{R}_i = .50$.

CHAPTER 11

Square Tables with Ordered Categories

The models and measures of association considered so far apply to classifications having arbitrary numbers of categories. In some applications, such as those where the same variable is measured for both members of a matched pair, the classifications in a table have the same categories. In social mobility tables, for example, each observation is a pairing of parent's social class with child's social class. In this chapter we consider specialized models and measures that can be applied to square tables having the same ordered row and column classifications.

There are two matters that are of special interest in the analysis of square tables. First, the cell probabilities may exhibit a type of symmetry in terms of their location relative to the main diagonal of the table. Section 11.1 presents models that describe various symmetry patterns and utilize the ordered nature of the rows and columns. Second, a comparison of the two marginal distributions is usually informative. For example, many studies concern whether a distribution shifts upward or downward after some event or period of time. Section 11.2 describes ways of comparing marginal distributions in square tables having ordered categories. In Section 11.3 some data are analyzed in which it is insufficient to apply only standard methods for square tables. This final example illustrates that it is often important to view data from different perspectives.

11.1 MODELS FOR SQUARE TABLES

Let m_{ij} be the expected frequency in the cell in row i and column j of a $r \times r$ table. The main diagonal contains the $\{m_{ii}\}$, for which the classification is the same on the row and column variables. Models for square tables describe various patterns for the $\{m_{ij}\}$ about the main diagonal.

Symmetry

A very specialized model is that of *symmetry*,

$$m_{ij} = m_{ji} \tag{11.1}$$

for all cells. Assuming that this model holds, the common ML estimate of m_{ij} and m_{ji} is

$$\hat{m}_{ij} = \hat{m}_{ji} = \frac{(n_{ij} + n_{ji})}{2}$$

The symmetry model can be expressed as a loglinear model (see Complement 11.1), and it has df $= r(r-1)/2$ for testing goodness of fit through X^2 or G^2.

The symmetry model is appropriate for ordinal or nominal classifications. The other models to be discussed in this section apply only to ordinal variables, since the way a cell is treated depends on its distance from the main diagonal.

Diagonals-Parameter Models

Goodman (1972, 1979b) formulated several methods in which diagonals that are equidistant from the main diagonal exhibit similar patterns for expected cell frequencies. His *diagonals-parameter symmetry* model has the form

$$m_{ij} = m_{ji}\delta_{j-i}, \qquad i < j \tag{11.2}$$

for a set of parameters $\{\delta_1, \ldots, \delta_{r-1}\}$. In other words, each expected frequency on the main diagonal that is k units above the main diagonal is the same multiple δ_k of the corresponding expected frequency on the diagonal that is k units below the main diagonal. The parameter δ_k is the odds that an observation falls in a cell (i, j) satisfying $j - i = k$ instead of in a cell satisfying $j - i = -k$, $k = 1, \ldots, r-1$. The odds depend only on the distance k between the diagonal containing the cell and the main diagonal. This model has $r - 1$ more parameters than the symmetry model, and it has df $= (r-1)(r-2)/2$ for testing goodness of fit. The symmetry model is the special case of (11.2) in which $\delta_1 = \cdots = \delta_{r-1} = 1$.

Let T_k denote the $2 \times (r-k)$ table constructed using the two diagonals that are k units from the main diagonal, $k = 1, \ldots, r-2$. For example, the first row of T_1 is $(n_{12}, n_{23}, \ldots, n_{r-1,r})$ and the second row is $(n_{21}, n_{32}, \ldots, n_{r,r-1})$. When model (11.2) holds, the expected values of these counts are such that each entry in the first row of T_k is the same multiple (δ_k) of the corresponding entry in the second row. Model (11.2) is therefore equivalent to independence for the expected frequencies in the tables $\{T_k, k = 1, \ldots, r-2\}$. The $\{\hat{m}_{ij}\}$ for model (11.2) can be obtained by computing expected frequency estimates for the independence model when it is applied to the $\{T_k\}$ tables. The estimate of δ_k is the common ratio of expected frequency estimates in the rows of T_k.

Conditional Symmetry Model

Bishop et al. (1975, pp. 285–286) and McCullagh (1978) discussed the model

$$m_{ij} = \delta m_{ji}, \qquad i < j \tag{11.3}$$

This model implies that for $i < j$,

$$P(X = i, Y = j \mid X < Y) = P(X = j, Y = i \mid X > Y)$$

where (X, Y) is selected at random according to the $\{m_{ij}\}$ distribution. Because of this property, it is referred to as a *conditional symmetry* model. This model is the special case of the diagonals-parameter symmetry model (11.2) with $\delta_1 = \cdots = \delta_{r-1} = \delta$. It has only one more parameter than the symmetry model (11.1), so its residual df $= (r+1)(r-2)/2$. The symmetry model corresponds to $\delta = 1$ in model (11.3).

Let

$$\hat{\delta} = \frac{\sum\sum_{i<j} n_{ij}}{\sum\sum_{i>j} n_{ij}}$$

Then the $\{\hat{m}_{ij}\}$ for model (11.3) are

$$\hat{m}_{ij} = \frac{\hat{\delta}(n_{ij} + n_{ji})}{(\hat{\delta} + 1)}, \qquad i < j$$

$$\hat{m}_{ij} = \frac{(n_{ij} + n_{ji})}{(\hat{\delta} + 1)}, \qquad i > j$$

Other special cases of the diagonals-parameter symmetry model are sometimes of interest. For example, if it is reasonable to assume an underlying bivariate normal distribution, then the model

$$m_{ij} = m_{ji}\,\delta^{j-i}$$

may fit well (see Complement 11.5). This model is the special case of model (11.2) with $\delta_k = \delta^k$, $k = 1, \ldots, r-1$, so it assumes a linear trend in the logs of the diagonals parameters.

Quasi-Uniform Association Model

Goodman (1979a) observed that regular loglinear models for ordinal variables sometimes fit square tables well when the cells on the main diagonal are ignored. He proposed a model having uniform local association for cells off that diagonal. This model is called a *quasi-uniform association* model.

Consider the sample cross-classification table having cell counts $\{n_{ij}^*\}$ in which the main diagonal entries in the original table are replaced by zeroes; that is, $n_{ij}^* = n_{ij}$ for $i \neq j$ and all $n_{ii}^* = 0$. Let m_{ij}^* denote the expected frequency corresponding to n_{ij}^*. If the quasi-uniform association model holds with local odds ratio θ, then the $\{m_{ij}^*\}$ have the same local odds ratios as obtained in the table

$$\begin{pmatrix} 0 & 1 & 1 & 1 & 1 & \cdots \\ 1 & 0 & \theta^2 & \theta^3 & \theta^4 & \cdots \\ 1 & \theta^2 & 0 & \theta^6 & \theta^8 & \cdots \\ 1 & \theta^3 & \theta^6 & 0 & \theta^{12} & \cdots \\ 1 & \theta^4 & \theta^8 & \theta^{12} & 0 & \cdots \end{pmatrix}$$

In fact the $\{m_{ij}^*\}$ equal these values rescaled (by iterative proportional fitting) to have the row marginal distribution $\{m_{i+}^*\}$ and the column marginal distribution $\{m_{+j}^*\}$.

The quasi-uniform association model has residual df $= r(r-3)$, so it is unsaturated only for tables of size 4×4 and larger. ML estimates $\{\hat{m}_{ij}^*\}$ of the $\{m_{ij}^*\}$ for this model can be obtained by applying to $\{n_{ij}^*\}$ the iterative scaling procedure described for the uniform association model in Section 5.1. Initial estimates for diagonal elements are taken to be zero, however, so that all diagonal elements in successive cycles also equal zero.

The quasi-uniform association model for the case $\theta = 1$ is referred to as the *quasi-independence* model. This has been perhaps the most commonly applied model for square tables. It has an additional degree of freedom (since θ is given), and it is also appropriate for square tables with unordered categories. If the quasi-independence model holds, then given that the classification on the row variable differs from the classification of the column variable, the variables are independent. This model can be fitted using iterative proportional fitting. The ML estimates of $\{m_{ij}^*\}$ are the limit of the sequence

$$m_{ij}^{(t+1)} = m_{ij}^{(t)}\left(\frac{n_{i+}^*}{m_{i+}^{(t)}}\right)$$

$$m_{ij}^{(t+2)} = m_{ij}^{(t+1)}\left(\frac{n_{+j}^*}{m_{+j}^{(t+1)}}\right)$$

$t \geq 0$, where $m_{ij}^{(0)} = 1$ for $i \neq j$ and $m_{ii}^{(0)} = 0$. See Note 4.10 for a generalization of this model.

Other Approaches

Haberman (1979, pp. 500–503) presented other loglinear models in which the effect of a cell on the association depends on its distance from the main diagonal. McCullagh (1977, 1978) suggested two models that are more easily described in terms of groupings of cells than individual cells. Consider the $r - 1$ ways of collapsing the $r \times r$ table into a 2×2 table by choosing cut points after the kth row and after the kth column, $k = 1, \ldots, r - 1$. McCullagh's logistic model and palindromic invariance model imply a constant ratio of the two off-diagonal cells for these $r - 1$ collapsings; that is,

$$\sum_{i \leqslant k} \sum_{j > k} m_{ij} = \delta \sum_{i > k} \sum_{j \leqslant k} m_{ij}, \qquad k = 1, \ldots, r - 1 \tag{10.4}$$

The palindromic invariance model is another generalization of the conditional symmetry model. The logistic model applies when one marginal distribution is a location shift of the other marginal distribution on a logistic scale; that is, when

$$\log\left[\frac{P(X \leq i)}{(1 - P(X \leq i))}\right] = \log\left[\frac{P(Y \leq i)}{(1 - P(Y \leq i))}\right] + \Delta, \qquad i = 1, \ldots, r - 1$$

Occupational Mobility Example

The data in Table 11.1, taken from Glass (1954, p. 183), have been analyzed by several statisticians, including Mosteller (1968), Goodman (1972), and Bishop et al. (1975, p. 100). The table relates father's and son's occupational status category for a British sample. There are 3500 observations, of which 2041 fall off the main diagonal. Except for the (n_{13}, n_{31}) combination, the sample satisfies $n_{ij} > n_{ji}$ whenever $i < j$. Thus for each pair of categories (i, j) with $i < j$ except for (1, 3), the proportion of the father-son pairs for which the son had the higher status category is greater than the proportion for which the father had the higher status category.

Table 11.1 contains estimated expected frequencies for the symmetry model (11.1), the conditional symmetry model (11.3), and the diagonals-parameter symmetry model (11.2). The symmetry model ($m_{ij} = m_{ji}$ for all cells) fits poorly, with $G^2 = 37.5$ based on df $= r(r - 1)/2 = 10$. The conditional symmetry model ($m_{ij} = \delta m_{ji}$ for $i < j$) has only one additional parameter but fits quite well, with $G^2 = 10.4$ based on df $= (r + 1)(r - 2)/2 = 9$. For this model $\hat{\delta} = 1.26$. It has the simple interpretation that for each pair of categories, the proportion of father-son pairs for which the son had the higher status is estimated to be 1.26 times higher than the proportion in which the father had the higher status.

The diagonals-parameter symmetry model ($m_{ij} = m_{ji} \delta_{j-i}$ for $i < j$) also fits well, with $G^2 = 6.4$ based on df $= (r - 1)(r - 2)/2 = 6$. The reduction in

Table 11.1. Occupational Status for British Father-Son Pairs

Father's Status	Son's Status 1	2	3	4	5	Total
1	50	45	8	18	8	129
	—	(36.5)[a]	(9.5)	(16.0)	(5.5)	
	—	(40.7)[b]	(10.6)	(17.8)	(6.1)	
	—	(41.4)[c]	(10.0)	(18.1)	(8.0)	
	—	(35.6)[d]	(16.1)	(21.8)	(5.5)	
2	28	174	84	154	55	495
	(36.5)	—	(81.0)	(152.0)	(48.5)	
	(32.3)	—	(90.3)	(169.5)	(54.1)	
	(31.6)	—	(91.8)	(159.7)	(54.9)	
	(25.6)	—	(83.6)	(157.3)	(54.5)	
3	11	78	110	223	96	518
	(9.5)	(81.0)	—	(204.0)	(84.0)	
	(8.4)	(71.7)	—	(227.5)	(93.7)	
	(9.0)	(70.2)	—	(231.2)	(88.3)	
	(12.5)	(89.9)	—	(206.0)	(99.5)	
4	14	150	185	714	447	1510
	(16.0)	(152.0)	(204.0)	—	(383.5)	
	(14.2)	(134.5)	(180.5)	—	(427.7)	
	(13.9)	(144.3)	(176.8)	—	(434.6)	
	(15.1)	(150.8)	(183.7)	—	(446.5)	
5	3	42	72	320	411	848
	(5.5)	(48.5)	(84.0)	(383.5)	—	
	(4.9)	(42.9)	(74.3)	(339.3)	—	
	(3.0)	(42.1)	(79.7)	(332.4)	—	
	(2.8)	(38.7)	(65.7)	(329.9)	—	
Total	106	489	459	1429	1017	3500

Source: Glass (1954). Reprinted by permission.
[a]Estimated expected frequencies for symmetry model.
[b]Estimated expected frequencies for conditional symmetry model.
[c]Estimated expected frequencies for diagonals-parameter symmetry model.
[d]Estimated expected frequencies for quasi-uniform association model.

G^2 from the conditional symmetry model is $10.4 - 6.4 = 4.0$ based on $\text{df} = 9 - 6 = 3$. This statistic tests H_0: $\delta_1 = \delta_2 = \delta_3 = \delta_4$, given that the diagonals-parameter symmetry model holds. The estimated odds values are $\hat{\delta}_1 = 1.31$, $\hat{\delta}_2 = 1.11$, $\hat{\delta}_3 = 1.30$, and $\hat{\delta}_4 = 2.67$. The fourth estimate is based on only 11 observations ($\hat{\delta}_4 \doteq n_{14}/n_{41} = 8/3$), and the differences among these values are not significant. Hence the conditional symmetry model represents these data adequately.

Table 11.1 also contains estimated expected frequencies for the quasi-uniform association model. This model fits fairly well, with $G^2 = 13.9$ based on df $= r(r-3) = 10$. For local 2×2 tables that do not contain a cell on the main diagonal, the odds that a son's status is $j+1$ instead of j is estimated to be $\hat{\theta} = 1.39$ times higher when the father's status is $i+1$ than when it is i. For $a < b$ and $c < d$ with $a \neq c$, $a \neq d$, $b \neq c$, $b \neq d$, the odds that a son's status is d instead of c is estimated to be exp $[1.39(b-a)(d-c)]$ times higher when the father's status is b than when it is a. Hence the sample exhibits a fairly substantial positive association between father's status and son's status. The simpler quasi-independence model ($\theta = 1$ off the main diagonal) fits very poorly, with $G^2 = 235.8$ based on df $= 11$. The large increase in the G^2 value provides very strong evidence that $\theta > 1$.

11.2. MARGINAL HOMOGENEITY

Next we consider the problem of comparing the two marginal distributions of a square table. The condition of marginal homogeneity states that $\pi_{i+} = \pi_{+i}$, $i = 1, \ldots, r$. Marginal homogeneity is implied by, but does not imply, symmetry. A test statistic for the null hypothesis of marginal homogeneity is constructed as follows. Denote the sample differences in the paired marginal proportions by $d_i = p_{i+} - p_{+i}$, $i = 1, \ldots, r$, and let $\mathbf{d}' = (d_1, \ldots, d_{r-1})$. It is redundant to include d_r in this vector, since it is determined by $(d_1, \ldots, d_{r-1})$ through the condition $\sum_{i=1}^{r} d_i = 0$.The test statistic is

$$Q = n\,\mathbf{d}'\hat{\mathbf{W}}^{-1}\mathbf{d} = n \sum_i \sum_j d_i d_j \hat{u}_{ij} \tag{11.5}$$

where $\hat{\mathbf{W}} = (\hat{w}_{ij})$ is the estimated covariance matrix of $\sqrt{n}\,\mathbf{d}$, and where $\hat{u}_{ij}$ is the element in row i and column j of the matrix $\hat{\mathbf{U}} = \hat{\mathbf{W}}^{-1}$. The elements of $\hat{\mathbf{W}}$ are $\hat{w}_{ij} = -(p_{ij} + p_{ji}) - (p_{i+} - p_{+i})(p_{j+} - p_{+j})$ for $i \neq j$ and $\hat{w}_{ii} = p_{i+} + p_{+i} - 2p_{ii} - (p_{i+} - p_{+i})^2$.

Under the null hypothesis of marginal homogeneity, Q has an asymptotic chi-squared distribution with df $= r - 1$. The statistic Q was suggested by Bhapkar (1966). Stuart (1955) proposed a statistic of similar form, but with $\hat{\mathbf{W}}$ replaced by the *null* estimated covariance matrix. Another chi-squared statistic having df $= r - 1$, based on ML estimates of expected frequencies satisfying marginal homogeneity, was given by Bishop et al. (1975, pp. 294–295).

Homogeneity in Ordinal Models

The statistic (11.5) can be applied to nominal or ordinal classifications. Other tests can be formulated that utilize ordinality. The conditional symmetry model, the logistic model, the palindromic invariance model, and the diagonals-parameter symmetry model with all $\delta_k \geq 1$ or all $\delta_k \leq 1$ imply that the marginal distributions are stochastically ordered. For each model, marginal homogeneity means that the model parameters equal one. Hence, given that any one of these models holds, a test of marginal homogeneity can be constructed by testing that the parameter(s) equal one.

To illustrate, suppose that the conditional symmetry model (11.3) fits well. The difference between the G^2 values for the symmetry model (11.1) and for model (11.3) has an asymptotic chi-squared distribution with df = 1 for testing H_0: $\delta = 1$. Note that if model (11.2) or model (11.3) holds, then marginal homogeneity is equivalent to symmetry. If one of the ordinal models fits the data well, tests of marginal homogeneity based on it will tend to be more powerful than tests [e.g., (11.5)] that ignore the ordering of the categories.

Mann-Whitney Test

It is also possible to compare the marginal distributions using general statistics for comparing ordinal categorical distributions. However, the assessment of sampling variability for such a statistic must take into account the matching (rather than independence) of the samples comprising the observed marginal distributions.

If a research hypothesis in a study predicts that one marginal distribution is stochastically higher than the other, a useful measure of marginal inhomogeneity is given by

$$\hat{\Delta} = \sum_{i<j}\sum p_{i+} p_{+j} - \sum_{i>j}\sum p_{i+} p_{+j} \tag{11.6}$$

This statistic is the difference between discrete analogs of the Mann-Whitney statistics. Its population value can be expressed as

$$\Delta = P(Y > X) - P(X > Y)$$

where X is selected at random from the $\{\pi_{i+}\}$ marginal distribution, and Y is selected independently at random from the $\{\pi_{+j}\}$ marginal distribution. This measure is positive when Y is stochastically higher than X and negative when X is stochastically higher than Y. Marginal homogeneity implies that $\Delta = 0$, but the converse is not true. For instance, Δ can equal 0 when the marginal distributions have similar locations but different variabilities.

The measure $\Delta = P(Y > X) - P(X > Y)$ was first defined in (9.10). There, however, it was estimated for a $2 \times c$ table having separate samples in the two rows. Statistic (11.6), by contrast, is computed for a $2 \times r$ table consisting of the marginal distributions of a $r \times r$ table, so that the samples in the two rows are actually matched.

Several alternative expressions can be given for Δ that utilize the equivalent ways of comparing distributions through Mann-Whitney statistics, Wilcoxon mean rank statistics, or mean ridit statistics. For example, let $R_U(V)$ denote the mean ridit for the distribution of V when the distribution of U is the identified distribution for calculating the ridits. In terms of the distribution functions $\{F_i^X\}$ and $\{F_i^Y\}$ of the marginal variables X and Y,

$$R_X(Y) = \sum \pi_{+j}\left[\frac{(F_j^X + F_{j-1}^X)}{2}\right], \qquad R_Y(X) = \sum \pi_{i+}\left[\frac{(F_i^Y + F_{i-1}^Y)}{2}\right]$$

and $R_X(X) = R_Y(Y) = 0.5$. It is easily seen that

$$\begin{aligned}\Delta &= R_X(Y) - R_Y(X)\\ &= 2(R_Y(Y) - R_Y(X)) = 2(R_X(Y) - R_X(X)) = 2R_X(Y) - 1 \qquad (11.7)\end{aligned}$$

The delta method implies that $\hat{\Delta}$ also has a large-sample normal distribution for this case in which the samples comprising the marginal distributions are matched. Assume that the proportions $\{p_{ij}\}$ in the square table result from full multinomial sampling, and let

$$\hat{\phi}_{ij} = \hat{F}_j^X + \hat{F}_{j-1}^X - \hat{F}_i^Y - \hat{F}_{i-1}^Y$$

where $\{\hat{F}_i^X\}$ and $\{\hat{F}_i^Y\}$ are the sample marginal distribution functions. For large n, it follows that $\hat{\Delta}$ is approximately normally distributed with mean Δ and with variance that can be estimated by

$$\hat{\sigma}^2(\hat{\Delta}) = \frac{\sum\sum \hat{\phi}_{ij}^2 p_{ij} - (\sum\sum \hat{\phi}_{ij} p_{ij})^2}{n}$$

For large n, therefore, the null hypothesis of marginal homogeneity can be tested using the statistic $z = \hat{\Delta}/\hat{\sigma}(\hat{\Delta})$, which has approximately a standard normal null distribution.

Unlike the chi-squared test (11.5), this z-test is not consistent for the class of all alternatives but only those for which $\Delta \neq 0$. The z-test can be much more powerful than the chi-squared test when the true marginal distributions are stochastically ordered. Some numerical results of Agresti (1983b) show that this is especially true when the number of ordered categories is large and the dependence is strong.

Occupational Mobility Example

Consider again the data in Table 11.1 on occupational status for 3500 British father-son pairs. For the population represented by this sample, one might analyze whether the occupational distribution for sons differs from the occupational distribution for fathers. The sample marginal distributions are stochastically ordered, and there is a marked tendency for more sons to have occupations with the highest status.

The vector of differences in sample marginal distributions is $\mathbf{d}' = (23/3500, 6/3500, 59/3500, 81/3500, -169/3500)$. Bhapkar's statistic (11.5) equals $Q = 32.9$ based on df $= 4$, which shows very strong evidence ($P < 0.001$) of a difference in the distributions.

Since the conditional symmetry model fits Table 11.1 well, marginal homogeneity can also be examined by testing H_0: $\delta = 1$ in that model. The test statistic is the difference between the G^2 values for the symmetry model and the conditional symmetry model, based on df $= 1$. For Table 11.1, this difference is $37.5 - 10.4 = 27.1$, giving overwhelming evidence of a difference between the occupational distributions. This test statistic has nearly as large a value here as the Q-statistic, but it uses the ordinal nature of the classifications and is based on only 1 instead of 4 degrees of freedom. The estimate $\hat{\delta} = 1.26$ indicates that the occupational status distribution is stochastically higher for sons than for fathers.

For the ordinal Mann-Whitney approach the sample $\hat{\Delta} = 0.0507$ also indicates a tendency for the son's occupational status to be higher than father's. The standard error is $\hat{\sigma}(\hat{\Delta}) = 0.0102$, so the z-statistic $z = 0.0507/0.0102 = 4.99$ can be used to test H_0: $\Delta = 0$. This approach also provides extremely strong evidence of a difference in the marginal distributions.

It is not surprising that the symmetry model was observed in the previous section to fit these data so poorly. Whenever there is marked marginal inhomogeneity, the symmetry model is doomed.

11.3. A FINAL EXAMPLE

In the examples given in this text the ordinal methods have worked well and produced unambiguous results. Hence those data sets were useful for illustrating the use and interpretation of ordinal methods. In practice, of course, data sets can be considerably more complex, it may not be possible to state simple and general conclusions, and a variety of analyses may be necessary to gain a good understanding of the data. The final example in this book illustrates a data set for which the methods of this chapter seem well suited, yet for which the use of those methods alone would give incomplete and even misleading conclusions.

Table 11.2 is a cross-classification of police classification of a homicide (P), court classification of a homicide (C), race of defendant and race of victim (R). The sample, described by Radelet and Pierce (1983), consists of 1017 individuals indicted for homicide in Florida between 1973 and 1977. Each case is classified in police records as a "felony," "possible felony," or "nonfelony." This refers to the police department's judgment about whether the homicide was committed concurrently with another felony, such as robbery or rape. The prosecutor, in subsequently presenting the case in court, also classifies the homicide in one of these three categories. The prosecutor may ignore felony evidence given by the police (downgrade the classification), or find felony evidence that the police department ignored or overlooked (upgrade the classification), or give the same classification as the police department.

Table 11.2. Police and Court Classifications of 1017 Homicides, by Races of Defendant and Victim

		Court Classification			
Race of Defendant/Victim	Police Classification	No Felony	Possible Felony	Felony	Total
Black/white	No felony	7	1	3	11
	Possible felony	0	2	6	8
	Felony	5	5	109	119
	Total	12	8	118	138
White/white	No felony	236	11	26	273
	Possible felony	7	2	21	30
	Felony	25	4	101	130
	Total	268	17	148	433
Black/black	No felony	328	6	13	347
	Possible felony	7	2	3	12
	Felony	21	1	36	58
	Total	356	9	52	417
White/black	No felony	14	1	0	15
	Possible felony	6	1	1	8
	Felony	1	0	5	6
	Total	21	2	6	29

Source: Radelet and Pierce (1983).

The two separate classifications of the homicides may be compared in terms of whether the court classifications tended to be more severe or less severe than the police classifications. The Radelet and Pierce study investigated whether the relative frequencies of upgrading and downgrading in the court classification (relative to the police classification) were dependent on the defendant/victim race classification. They considered, for example, whether upgrading was more likely to occur when a black killed a white (BkW) than when a white killed a white (WkW). Radelet and Pierce were more concerned with making pairwise comparisons of the four defendant/victim classes than with assessing separate effects due to defendant's race or victim's race. Hence we shall treat the four defendant/victim race combinations as levels of a single variable "race," with levels BkW, WkW, BkB, WkB.

At each of the four race levels there is a square (3×3) table relating court classification and police classification. In each table most of the observations fall on the main diagonal, indicating an agreement in the two classifications. A striking feature of Table 11.2 is the symmetric pattern of the cell frequencies about the main diagonal in each table. The only major discrepancy is the (felony, possible felony) pair of cells in the WkW table. The WkB table has very small cell frequencies, so it is risky to search for patterns in it. The G^2 statistics for testing symmetry [model (11.1)] in the 3×3 tables are $G^2(\text{BkW}) = 1.98$, $G^2(\text{WkW}) = 13.59$, $G^2(\text{BkB}) = 3.02$, and $G^2(\text{WkB}) = 6.74$, each having df $= 3$. The overall G^2 for testing the model that symmetry holds within each race combination is the sum of these, $G^2 = 25.33$, based on df $= 12$. The contribution to it from the (felony, possible felony) pair of cells in the WkW table is 12.67, so the symmetry model fits fairly well over the remaining cells.

For each 3×3 table the sample marginal distributions are quite similar, with the discrepancy in the WkW table explained by the (felony, possible felony) pair of cells. With the exception of those two cells, there is not a marked tendency for prosecutors to upgrade the police classifications more frequently or to a greater extent than they downgrade them. Hence it is tempting to conclude that upgrading and downgrading of the police classification by prosecutors is independent of the race classification.

There is one important aspect of Table 11.2 not yet discussed, however. For the BkW table nearly all police classifications are "felony," whereas for WkW and especially BkB the overwhelming majority of police classifications are "no felony." In general, the marginal distributions of police classifications are vastly different for the four race groups. There was not much potential for upgrading to occur with the BkW data, whereas there was considerable potential for the WkW and BkB data. Therefore it is necessary to control for police classification in comparing the relative frequencies of upgrading and downgrading for these racial groups.

In fact for a given police classification there are noticeable differences in the court classifications among the four race groups. For example, for those cases classified felony by the police, the court classification is also felony 91.6 percent (109/119 = 0.916) of the time for BkW but only 62.1 percent of the time for BkB. We shall now consider a loglinear model for which the race groups can be compared with respect to court classification, controlling for police classification.

Let m_{ijk} be the expected frequency at level i of R, level j of P, and level k of C, for the $4 \times 3 \times 3$ table. According to the loglinear model

$$\log m_{ijk} = \mu + \lambda_i^R + \lambda_j^P + \lambda_k^C + \lambda_{ij}^{RP} + \lambda_{jk}^{PC} \quad (11.8)$$

court classification is independent of racial group, given police classification. This model fits very poorly, with $G^2 = 73.75$ based on df = 18. Next consider the model

$$\log m_{ijk} = \mu + \lambda_i^R + \lambda_j^P + \lambda_k^C + \lambda_{ij}^{RP} + \lambda_{jk}^{PC} + \tau_i^{RC}(w_k - \bar{w}) \quad (11.9)$$

in which race effects $\{\tau_i^{RC}\}$ satisfying $\sum \tau_i^{RC} = 0$ are introduced for a R–C association. This is a "row effects" model for the conditional R–C association, in which the variable C is treated as ordinal through monotonic scores $\{w_k\}$ for its levels. For integer scores this model fits quite well, with $G^2 = 16.88$ based on df = 15. Model (11.9) has only three more independent parameters than the R–C conditional independence model (11.8), which is the special case of (11.9) with all $\tau_i^{RC} = 0$.

The reduction in G^2 of $73.75 - 16.88 = 56.87$ based on df = 3 provides very strong evidence that the court classification varies by race, controlling for police classification. The R–C association can be interpreted using the ML estimates of the race effects, as described for the row effects model (5.5). These estimates (with asymptotic standard errors in parentheses) are

$$\hat{\tau}_{\text{BkW}} = 0.808\ (0.118)$$

$$\hat{\tau}_{\text{WkW}} = 0.127\ (0.080)$$

$$\hat{\tau}_{\text{BkB}} = -0.417\ (0.087)$$

$$\hat{\tau}_{\text{WkB}} = -0.517\ (0.282)$$

For example, $\exp(\hat{\tau}_{\text{BkW}} - \hat{\tau}_{\text{BkB}}) = 3.4$. Given police classification, the ratio of cases of felony to possible felony, and the ratio of cases of possible felony to no felony for the court classification is estimated to be 3.4 times higher for BkW than for BkB.

When the computer packages GLIM or SPSS^{X} (LOGLINEAR) are used to fit model (11.9), the covariance matrix of the parameter estimates can be obtained. From this matrix one can calculate the standard error of a difference of $\hat{\tau}$ values and make inferences about differences between pairs of racial groups. This analysis gives strong evidence of a difference for each pair except (BkB, WkB). Unlike the square tables symmetry analysis, the analysis using model (11.9) reveals rather large racial effects in the prosecutor's handling of police classifications. These effects can be described in a more complete manner by fitting a log-multiplicative model for the R–C association in which the scores for C are parameters. See Complement 11.4.

NOTE

Section 11.3

In the example the combinations of defendant's race and victim's race were treated as four levels of a single variable (race). This means that model association terms for race with another variable are actually three-factor interaction terms for defendant's race, victim's race, and that variable.

COMPLEMENTS

1. Consider the loglinear model

$$\log m_{ij} = \mu + \lambda_i + \lambda_j + \lambda_{ij}^{XY}$$

where $\sum \lambda_i = \sum_i \lambda_{ij}^{XY} = 0$ and all $\lambda_{ij}^{XY} = \lambda_{ji}^{XY}$.

   a. Show that this is the symmetry model.
   b. Suppose that $\lambda_i + \lambda_j$ were replaced in this model by $\lambda_i^X + \lambda_j^Y$, where $\sum \lambda_i^X = \sum \lambda_j^Y = 0$. The resulting model is called the *quasi-symmetry* model and has df $= (r-1)(r-2)/2$. Show that quasi symmetry plus marginal homogeneity equals symmetry. See Bishop et al. (1975, pp. 286–287) and McCullagh (1982). Goodman (1979a) referred to the quasi-symmetry model as the *symmetric association* model because it can be characterized by the property $\theta_{ij} = \theta_{ji}$ for all the local odds ratios.
   c. Show that the quasi-symmetry model fits Table 11.1 well, with $G^2 = 4.67$ based on df $= 6$.

2. The following data describe unaided distance vision for a sample of women.

Right Eye Grade	Left Eye Grade: Best	Second	Third	Worst
Best	1520	266	124	66
Second	234	1512	432	78
Third	117	362	1772	205
Worst	36	82	179	492

Source: Stuart (1955). Reprinted by permission of the Biometrika Trustees.

a. Fit the symmetry model to these data.

b. Fit the diagonals-parameter symmetry model, and interpret the fit. See Goodman (1979b).

c. Given that the model in (b) fits, show how to test for marginal homogeneity.

d. Use the measure (11.6) to test for marginal homogeneity. Interpret the result.

3. The accompanying table, taken from Glass (1954), was analyzed by Goodman (1979a). Table 11.1 is obtained from this table by combining categories 2 and 3 and categories 6 and 7.

Father's Status	Son's Status: 1	2	3	4	5	6	7
1	50	19	26	8	18	6	2
2	16	40	34	18	31	8	3
3	12	35	65	66	123	23	21
4	11	20	58	110	223	64	32
5	14	36	114	185	714	258	189
6	0	6	19	40	179	143	71
7	0	3	14	32	141	91	106

Source: Glass (1954). Reprinted by permission.

a. Show that the quasi-uniform association model fits these data well, with $G^2 = 30.4$ based on df $= 28$ and $\hat{\theta} = 1.26$.

b. Show that the quasi-independence model fits very poorly, with $G^2 = 408.4$ based on df $= 29$.

4. For Table 11.2, consider the model

$$\log m_{ijk} = \mu + \lambda_i^R + \lambda_j^P + \lambda_k^C + \lambda_{ij}^{RP} + \lambda_{jk}^{PC} + \beta\mu_i\omega_k$$

where the $\{\mu_i\}$ and $\{\omega_k\}$ are score parameters satisfying

$$\sum \mu_i = \sum \omega_k = 0 \quad \text{and} \quad \sum \mu_i^2 = \sum \omega_k^2 = 1$$

a. Show that this is a conditional association version of the log-multiplicative row and column effects model (8.1).

b. Fit this model to Table 11.2. Use the estimated scores to describe the conditional R–C association.

c. Compare the fit of this model to the fit of model (11.9). How does your interpretation of the R–C association differ with this model?

d. Note that the $\{\hat{\omega}_k\}$ for this model suggest that the felony and possible felony levels of court classification can be combined. Formulate and fit a corresponding logit model for this dichotomous definition of court classification.

5. Suppose that (X, Y) has density $f(x, y)$ of bivariate normal form, with $E(X) = \mu$, $E(Y) = \mu + \Delta$, Var (X) = Var $(Y) = \sigma^2$, and Corr $(X, Y) = \rho$.

a. Show that $f(x, y)/f(y, x)$ has the form δ^{x-y} for some constant δ.

b. If it is reasonable to assume an underlying normal distribution, argue that the model

$$m_{ij} = \delta^{j-i} m_{ji}$$

may be appropriate for a square ordinal table. (*Note*: This model is the special case of the diagonals-parameter symmetry model in which $\delta_k = \delta^k$. Its residual df is one less than for the symmetry model, which is the special case of this model with $\delta = 1$. See Agresti 1983c).

c. Show that the model in (b) can be expressed in loglinear form as

$$\log m_{ij} = \mu + \lambda_i + \lambda_j + \lambda_{ij}^{XY} + (j - i)\beta$$

where $\sum \lambda_i = \sum_i \lambda_{ij}^{XY} = 0$, all $\lambda_{ji}^{XY} = \lambda_{ij}^{XY}$, and $2\beta = \log \delta$.

d. Show that the model in (b) can also be expressed as the no-intercept logit model

$$\log\left(\frac{m_{ij}}{m_{ji}}\right) = \delta'(j - i), \quad \text{where } \delta' = \log \delta$$

e. Show that the model in (b) is also a special case of the quasi-symmetry model described in Complement 11.1(b).

f. In the accompanying table, taken from Breslow (1982), 80 esophageal cancer patients are compared with controls on the number of beverages drunk at "burning hot" temperatures. Fit the model in (b) to these data, and interpret the results.

	Controls			
Case	0	1	2	3
0	31	5	5	0
1	12	1	0	0
2	14	1	2	1
3	6	1	1	0

Source: Breslow (1982). Reprinted with permission from The Biometric Society.

CHAPTER 12

Comparison of Ordinal Methods

This book has outlined several methods for analyzing cross-classifications containing ordinal variables. In this chapter the primary methods are summarized and compared. In Section 12.1 we review methods for three settings: ordinal-ordinal tables, ordinal-nominal tables, and multidimensional tables in which the response variable is ordinal. The methods are compared in terms of the types of conclusions they allow us to make for the dumping severity and the political ideology data sets. Section 12.2 contains a fundamental comparison of the strategies. The discussion focuses on the advantages and disadvantages of the model-building approaches and the measures of association (model-free) approach. Specific comparisons are made of the loglinear and logit approaches for building models.

12.1. SUMMARY OF METHODS AND RESULTS

We have considered essentially three types of methods for describing contingency tables having ordinal classifications: models that do not require distinguishing between response and explanatory variables, models that do require such a distinction, and measures of association. In Chapters 3 through 8 we discussed model-building approaches in which the table structure was represented by descriptions such as "uniform association." The logit models in Chapter 6 and 7, unlike the loglinear models in Chapters 3 through 5, identify a response variable. The two types of models imply different association patterns when the number of response categories exceeds two. The third type of method, discussed in Chapters 9 and 10, involves summarizing the data through association measures such as gamma and tau-*b* for ordinal-ordinal tables and mean ridits for ordinal-nominal tables. These summary measures describe certain aspects of the data (e.g., relative numbers of concordant and discordant pairs), but they need not be model based, and they do not fully describe the structural pattern of the cell proportions.

Ordinal-Ordinal Tables

The primary methods we considered for ordinal-ordinal tables were as follows:

1. Loglinear models, such as the uniform association model [(5.1) with integer scores].
2. Logit models, such as the uniform association model for cumulative logits [(7.3) with integer scores].
3. Ordinal-ordinal measures of association, such as the concordance-based measures gamma, Kendall's tau-*b*, and Somers' *d*.

For each method there is a single degree of freedom chi-squared statistic for testing independence against an alternative that corresponds to nonnull values of an association parameter. For loglinear and logit models such a statistic is the difference between the G^2 values for the independence model and the uniform association model. The square of statistic (10.1) is an example of a chi-squared statistic designed to detect nonzero values for a class of measures of association. The alternatives in these tests are narrower than the broad alternative of dependence, and the measures for which the tests are formulated give various ways of characterizing monotonicity of the association.

For the cross-classification of dumping severity and operation, all methods give fairly strong evidence of an association: $G^2(I \mid U) = 6.29$ with $\text{df} = 1$ for the difference in G^2 values between the independence model and the loglinear uniform association model; $G^2(I \mid U) = 6.61$ with $\text{df} = 1$ for the difference in G^2 values between the independence model and logit uniform association model; and $z = 2.55$ ($z^2 = 6.50$, $\text{df} = 1$) for the ordinal measures approach. The loglinear and logit uniform association models both fit the data quite well, with $G^2 = 4.59$ and $G^2 = 4.27$ respectively, both based on $\text{df} = 5$. They also produced very similar estimates of expected frequencies, as shown by Tables 5.1 and 7.1. The association parameter estimates obtained with all methods indicate that the sample association is positive but relatively weak: $\hat{\gamma} = 0.170$, $\hat{\tau}_b = 0.111$, $d_{YX} = 0.096$ for the ordinal measures, $\hat{\beta} = 0.163$ for the constant local log odds ratio implied by the loglinear uniform association model, and $\hat{\beta} = 0.225$ for the constant local-global log odds ratio implied by the logit uniform association model. Semenya et al. (1983) applied several alternative models to these data. Similar substantive results are obtained using their models or other ordinal methods described in this text, such as the global odds ratio and the mean response models described in Chapter 8.

In practice, the association parameter estimates for the loglinear and cumulative logit models are usually not the same size, since they estimate different parameters. Specifically, the estimates refer to different log odds ratios. For the loglinear uniform association model, $\hat{\beta}$ is the constant predicted log odds ratio for the $(r-1)(c-1)$ 2×2 tables formed using only adjacent rows and adjacent columns. For the cumulative logit uniform association model, $\hat{\beta}$ is the constant predicted log odds ratio for the $(r-1)(c-1)$ 2×2 tables formed using adjacent levels of X but the various collapsings of the response into two categories. Figures 5.1 and 7.1 illustrate these two types of constant odds ratios. When both models fit fairly well (as with the dumping severity data), $|\hat{\beta}|$ for the loglinear model is typically smaller than $|\hat{\beta}|$ for the logit model. In other words, local associations tend to be weaker than associations that are global in one of the dimensions. See Note 7.2. Similarly, both of these values tend to be smaller than the absolute value of the association parameter estimate in the uniform association model (8.5) for global odds ratios. For that model the estimated log odds ratio is 0.44 for the dumping severity data.

Ordinal-Nominal Tables

The primary methods we considered for ordinal-nominal tables were as follows:

1. Loglinear models, such as the row effects model (5.5).
2. Logit models, such as the cumulative logit model (7.6) having homogeneous row effects.
3. Ordinal-nominal measures of association and location, such as delta, alpha, and the mean ridits.

For each method there is a chi-squared test with $\text{df} = r - 1$ (where r is the number of categories of the nominal variable) for testing independence against an alternative that corresponds to nonnull values of the association parameters. For example, see (5.7) and (7.8). The alternatives give various ways of characterizing a stochastic ordering of the levels of the nominal variable with respect to their conditional distributions on the ordinal variable.

For the cross-classification of political ideology and party affiliation, all methods provide very strong evidence of an association: $G^2(I \mid R) = 102.85$

with df = 2 for the difference in G^2 values between the independence model and the loglinear row effects model; $G^2(I \mid R) = 100.96$ with df = 2 for the difference in G^2 values between the independence model and the cumulative logit row effects model; and $W = 96.6$ with df = 2 for the Kruskal-Wallis comparison of mean ridits. The loglinear and logit row effects models both fit the data fairly well, as $G^2 = 2.81$ and $G^2 = 4.70$ respectively, both with df = 2. Comparing Tables 5.3 and 7.2, we see that they also produce very similar estimates of expected frequencies. All association parameter estimates indicate that, in the sample, Democrats (row 1) were the least conservative and Republicans (row 3) were much more conservative than the other two affiliations. For example, the estimated model row effects (with standard errors in parentheses) are $\hat{\tau}_1 = -0.495$ (0.062), $\hat{\tau}_2 = -0.224$ (0.059), and $\hat{\tau}_3 = 0.719$ (0.080) for the loglinear model and $\hat{\tau}_1 = -0.670$ (0.083), $\hat{\tau}_2 = -0.282$ (0.079), and $\hat{\tau}_3 = 0.952$ (0.102) for the cumulative logit model. Similar substantive conclusions are obtained using other ordinal methods described in this text, such as the log-multiplicative model (8.1).

For the loglinear model $\hat{\tau}_b - \hat{\tau}_a$ is the constant predicted log odds ratio for the $(c-1)$ 2×2 tables formed using only adjacent columns of the ordinal variable in rows a and b. For the cumulative logit model $\hat{\tau}_b - \hat{\tau}_a$ is the constant predicted log odds ratio for the $c - 1$ possible ways of collapsing the ordinal variable into a dichotomy. When both models fit well (as with the political ideology data), $|\hat{\tau}_b - \hat{\tau}_a|$ tends to be smaller for the loglinear model, for which associations are local in the dimension of the ordinal variable.

Multidimensional Tables

The primary methods we considered for multidimensional tables were as follows:

1. Loglinear models in which ordinal variables contribute to the association and interaction terms through linear departures from the null conditions. For example, see (5.11), (5.13), (5.15), (5.17), and (5.19).
2. Logit models in which ordinal explanatory variables have homogeneous, linear effects on the logits for an ordinal response variable. See Table 7.3, for example.
3. Measures of partial association obtained by taking weighted averages of bivariate ordinal measures of association computed at the combinations of levels of control variables. For example, see (9.15).

The methods for multidimensional tables were illustrated using the three-dimensional cross-classification of dumping severity, operation, and hospital originally given in Table 4.9. All ordinal methods provide strong evidence of a positive association between dumping severity and operation, and they all indicate that the association can be assumed to have the same strength for each hospital. It can be described, for instance, by the common log odds ratios that were implied by the loglinear and logit uniform association models for the marginal O–D table, or by the ordinal partial measures $\hat{\gamma}_{OD \cdot H} = 0.184$ and $\hat{\tau}_{OD \cdot H} = 0.118$. Also the loglinear and logit analyses showed that, for each operation, dumping severity and hospital can be treated as conditionally independent.

12.2. COMPARISON OF STRATEGIES

For the examples we have considered, there is a remarkable similarity in the substantive results of the various ordinal analyses. Whether we used loglinear models, logit models, other types of ordinal models, or measures of association, we reached similar conclusions about whether variables are associated and about the basic nature of those associations.

Ordinal vs. Nominal Treatment of Data

The association parameters in the ordinal methods describe ordinal characteristics of the data, such as monotonicity and stochastic orderings. Hence descriptive statements made with these methods are almost always more informative than those based on methods that ignore the ordinal nature of the variables. In addition substantive inferential conclusions made using ordinal methods can differ considerably from those based on a strictly qualitative treatment of the data. We observed this fact with the data in Table 4.9, for which standard nominal loglinear analyses revealed little evidence of association between dumping severity and operation. Generally, when the methodology recognizes inherent ordinality, there is greater power for detecting certain forms of association, and there is a greater variety of ways of describing the association.

Because of these considerations the main theme of this book has been the importance of treating ordinal variables in a quantitative manner. The actual choice of ordinal method typically does not affect the substantive results. However, there are basic differences in emphasis in these methods that the reader should consider before selecting a strategy for analyzing ordinal data.

Measures of Association vs. Model Building

Logit and loglinear models summarize the structure of a table in such a way that, given model parameters, expected cell frequencies can be obtained. Therefore, if a particular model holds, there is no loss of information in using the model to represent that table. Measures of association, on the other hand, generally transform the cell proportions to a single number that describes a certain aspect of the association in the table. Thus a measure of association can be informative in giving us an indication of the strength of association in the table, but it does not give us as much information as a model. Models are especially useful for analyzing multidimensional tables, where it is more difficult using measures of association to describe parsimoniously the various possible association and interaction patterns. The models give a structure in which various conditions (such as conditional independence or no three-factor interaction) result when certain parameters in the model are equated to zero.

It can, however, be helpful to use measures of association as well as models in analyzing data. Most measures of association are quite simple to interpret, and they can be understood by a wide audience. Even if a simple model fits a table well, it usually enhances our understanding of the degree of association if we are given statistics such as the relative numbers of concordant and discordant pairs. Also for some cross-classifications none of the commonly used models provides an adequate fit, even if levels of the explanatory variables are stochastically ordered (see Agresti 1981a). Measures of association that compare stochastically ordered distributions can be used whenever the loglinear row effects, logit row effects, or various other models are appropriate. If different model types fit different tables, measures of association give us a common basis for comparing associations and summarizing the results of the models.

In summary, measures of association are quite widely applicable, but they give us less information about tables than do well-fitting models. Usually our primary goal will be to build a model that represents well the association patterns in a table. However, the use of measures of association can help us to understand the data and, for small tables, can be as illuminating as the use of models. In fact the two strategies are not mutually exclusive. In Section 10.5 we noted that interesting models can be formulated in terms of measures of association.

Logit vs. Loglinear Models

When there are only two response categories, the ordinal logit models are special cases of corresponding ordinal loglinear models. The association parameters in the logit models then equal those in the corresponding loglinear models when the two response scores are one unit apart in the loglinear model. For example, for a $r \times 2$ table the association parameter in the loglinear uniform association model and in the logit uniform association model equals the common log odds ratio for adjacent rows in the table. The equivalence in parameters does not occur when the number of response categories exceeds two,.and neither model is then a special case of the other.

On structural grounds neither model type is clearly preferable. The implications for odds ratio behavior seem reasonable for both. We obtained very similar results for the two types of models when we fit them to the dumping severity data and to the political ideology data. This similarity of results happens often in practice, as both model types imply that levels of explanatory variables are stochastically ordered on ordinal response variables.

In fact the two model types hold naturally for underlying continuous distributions of similar form, the logistic and the normal. For example, the cumulative logit row effects model (7.6) applies if the response variable has an underlying logistic distribution, since the logit transformation of the logistic distribution function $F(x) = [1 + \exp(-(x - \tau))]^{-1}$ is an additive function of the location parameter τ. The loglinear row effects model (5.5) applies if the underlying distribution of the ordinal variable is normal and if appropriate column scores are used. To be specific, suppose that two rows have underlying normal densities f_1 and f_2 with means μ_1 and μ_2 and the same variance σ^2. Then the relationship

$$\log\left[\frac{f_1(v_1)f_2(v_2)}{f_1(v_2)f_2(v_1)}\right] = \frac{(\mu_1 - \mu_2)(v_1 - v_2)}{\sigma^2}$$

is analogous to property (5.6) for the loglinear row effects model, when we identify μ_1/σ^2 and μ_2/σ^2 with τ_1 and τ_2. See also Goodman (1981b) and Complement (8.1) for correspondences between normal and loglinear models. Due to the similarity of the logistic and normal distributions, one would expect that if a loglinear model fits well, then so would the analogous cumulative logit model, and vice versa.

McCullagh (1980, pp. 121–122) argued that the cumulative logit model is preferable to the loglinear model because it is easier with the logit model to state conclusions about association parameters without reference to the groupings of response categories. If the logit model holds for the true cell proportions, then the model still holds, and the association parameters do not change values, if response categories are combined. If a loglinear model holds for the true cell proportions, on the other hand, it may not hold when categories of any variable are combined. Even if it does hold, the association parameters will not be on the same order of magnitude, since odds ratios will describe associations that are less local (more aggregated) than before.

In practice, this is probably not a serious deficiency of the loglinear model, since (1) the real underlying distributions are unlikely to be so nice that *either* model fits for all categorizations and (2) scoring systems can be used to make the association parameters in the loglinear model less dependent on the choice of categories. For example, scores can be defined to have means of zero and standard deviations of one with respect to the marginal distributions. For the linear-by-linear association model (5.1), the $\{u_i\}$ and $\{v_j\}$ are linearly rescaled to satisfy

$$\sum u_i \pi_{i+} = \sum v_j \pi_{+j} = 0 \quad \text{and} \quad \sum u_i^2 \pi_{i+} = \sum v_j^2 \pi_{+j} = 1$$

Then the adjusted association parameter β represents a constant log odds ratio for distances of one standard deviation on both variables.

To illustrate these points, suppose that the "slight" and "moderate" categories of dumping severity are combined in Tables 5.1 and 7.1, so that dumping severity is measured as "none" or "some." For the resulting 4×2 table the maximum likelihood estimate of the association parameter in the loglinear uniform association model and the cumulative logit uniform association model (which are identical since $c = 2$) is 0.229. As expected, this is similar to the value of $\hat{\beta} = 0.225$ for the cumulative logit uniform association model applied to the original table, but it is not similar to $\hat{\beta} = 0.163$ for the loglinear uniform association model. If we use standardized scores in the loglinear model, however, we get $\hat{\beta} = 0.124$ for the original table and $\hat{\beta} = 0.125$ for the collapsed table.

An advantage of the logit model is not requiring the assignment of scores to levels of a response variable that is inherently just ordinal in scale. Even with the logit model, though, scores must be assigned to levels of ordinal explanatory variables. In addition for both model types it is not really necessary to treat these scores as reasonable scalings of the ordinal variables in order for the models to be valid. For example, consider loglinear model (5.1). The row and column scores indicate how far apart the rows and the columns must be judged to be in order for the association to be linear by linear. If the model fits when a particular pair of row scores are relatively close, this tells us that the association is relatively weaker in that part of the table. This is useful information even if we do not truly believe that those two rows would be close together for an underlying interval-scale measurement of that variable.

The same remarks apply to the parameter scores in the log-multiplicative model (8.1) that has similar form to loglinear model (5.1). Underlying scales have monotone scores, whereas scores for model (8.1) tend to be nonmonotonic when associations change direction over various parts of the table.

Thus we can use such scores to provide information about the nature of the association without needing to regard them as indexes of how far apart the ordered levels truly are.

In summary, both the logit and loglinear models provide sensible strategies for analyzing ordinal categorical data. The main issue that influences the choice of model is whether it is important to identify a response variable. We have seen that for multidimensional tables it is easier to formulate the logit model, since it is unnecessary to consider association patterns among the explanatory variables. Since distinctions are usually made between response and explanatory variables, the logit models may be more useful in practice.

Other Models

Of the models discussed in this text, primary consideration has been given to the loglinear and logit strategies. These are the two most popular model types for standard categorical analyses (for nominal variables), and we believe that their ordinal formulations provide appealing ways of analyzing ordinal data. Nevertheless, other types of models can be more informative for certain purposes.

Log-multiplicative models such as (8.1) are useful if we are attracted to the basic form of ordinal loglinear models but prefer to treat the scores as parameters. However, log-multiplicative models have their own disadvantages. They are less parsimonious than loglinear models, and for multidimensional tables much more effort is required to obtain parameter estimates and associated covariance matrices.

Global odds ratio models [such as (8.5)] are useful when it is more natural to describe association through global than local associations. If an underlying continuous distribution is in Plackett's (1965) family of distributions having uniform global odds ratio, then model (8.5) holds for *all* ways of categorizing the variables. Hence the association parameter may be less dependent in value on choices of categorization and scoring than those in loglinear or logit models. The global odds ratio models seem to be less informative, however, for ordinal-nominal tables and for multidimensional tables having a single response variable.

Mean response models such as (8.8) are ordinal response analogs of standard regression and analysis of variance models for continuous response variables. However, these models do not represent the categorical structure of the data as well as do models for expected cell frequencies, and conditions such as independence do not occur as special cases of them.

Some models are more appropriate than loglinear or logit models when there is reason to assume a certain form for underlying response distributions. For instance, the proportional hazards model (8.10) is useful for underlying distributions that are location shifts on the log-log scale for the complement of the distribution function. Specialized models for cell behavior about the main diagonal are often useful for square tables. Finally, there are methods not considered in this book, such as the graphical method called correspondence analysis (see Benzécri 1976), that provide yet alternative views of the data.

In summary, there is a large variety of methods that can be used to analyze ordinal categorical data. We hope that the presentation of these methods in this book will enable researchers to perform meaningful and informative analyses.

APPENDIX A

Weighted Least Squares

In the examples in the text model fitting was generally done with the maximum likelihood (ML) method. The weighted least squares (WLS) method can also be used for fitting these models. Familiarity with the WLS method is useful for several reasons:

1. WLS computations have a standard form. Readers who do not have access to some of the computer packages described in Appendix D may be able to use (or write) a general program for WLS estimation in order to fit ordinal models.
2. WLS estimation is more readily adaptable than ML estimation for fitting models in which the response function is a complex function of the cell counts (e.g., a measure of association for a bivariate response).
3. WLS and ML estimates for many models are asymptotically (as $n \to \infty$) equivalent, in the sense that both are in the class of "best asymptotically normal" (BAN) estimates. For large samples the estimators are approximately normally distributed around the parameter value, and the ratio of the variances converges to one.
4. ML estimation often corresponds to iterative use of WLS. We shall observe this for the Newton-Raphson method of obtaining ML estimates, discussed in Appendix B.1.

We give a brief introduction to the WLS method in this appendix. A general discussion of the use of WLS for fitting linear models to categorical data was given by Grizzle, Starmer, and Koch (1969). For a detailed explanation of the use of WLS methods for fitting models to ordinal data, see the survey papers by Semenya and Koch (1980) and by Semenya et al. (1983).

A.1. WLS FOR CATEGORICAL DATA

Suppose that a response variable Y has c categories, and suppose that independent multinomial samples of sizes $n_1, \ldots, n_s$ are selected at the s levels of an explanatory variable, or at s combinations of levels of several explanatory variables. If the sampling is full multinomial, the analysis described here is conducted conditional on the cell counts $(n_1, \ldots, n_s)$ that constitute the joint marginal distribution of the explanatory variables. If the sampling is full multinomial and no response variable is identified, then $s = 1$ and c is the number of cells in the cross-classification table.

$$\text{Let } \boldsymbol{\pi}' = (\boldsymbol{\pi}'_{(1)}, \ldots, \boldsymbol{\pi}'_{(s)}), \quad \text{where} \quad \boldsymbol{\pi}'_{(i)} = (\pi_{1(i)}, \ldots, \pi_{c(i)})$$

denotes the population conditional probabilities at level i of the explanatory variables. Let $\mathbf{p}$ denote the sample proportions corresponding to $\boldsymbol{\pi}$, and let $\mathbf{V}(\boldsymbol{\pi})$ be the $cs \times cs$ covariance matrix of $\mathbf{p}$, that is,

$$\mathbf{V}(\boldsymbol{\pi}) = \begin{pmatrix} \mathbf{V}_1 & & 0 \\ & \ddots & \\ 0 & & \mathbf{V}_s \end{pmatrix}$$

where

$$n_i \mathbf{V}_i = \begin{pmatrix} \pi_{1(i)}(1 - \pi_{1(i)}) & -\pi_{1(i)}\pi_{2(i)} & \cdots & -\pi_{1(i)}\pi_{c(i)} \\ -\pi_{2(i)}\pi_{1(i)} & \pi_{2(i)}(1 - \pi_{2(i)}) & \cdots & -\pi_{2(i)}\pi_{c(i)} \\ \vdots & \vdots & & \vdots \\ -\pi_{c(i)}\pi_{1(i)} & -\pi_{c(i)}\pi_{2(i)} & \cdots & \pi_{c(i)}(1 - \pi_{c(i)}) \end{pmatrix}$$

Now, the response functions, denoted by $f_m(\boldsymbol{\pi})$, $m = 1, \ldots, u \leq s(c-1)$, are functions of $\boldsymbol{\pi}$ that are assumed to have continuous second-order partial derivatives. Let

$$\mathbf{F}(\boldsymbol{\pi})' = [f_1(\boldsymbol{\pi}), \ldots, f_u(\boldsymbol{\pi})]$$

and let $\mathbf{Q}(\boldsymbol{\pi})$ be the $u \times cs$ matrix

$$\mathbf{Q}(\boldsymbol{\pi}) = \left(\frac{\partial f_m(\boldsymbol{\pi})}{\partial \pi_{j(i)}} \right)$$

for $m = 1, \ldots, u$ and for all cs combinations (i, j). We assume that the $\{f_m\}$ are linearly independent, so that Q has rank u. Linear response functions have the form $\mathbf{F}(\boldsymbol{\pi}) = \mathbf{A}\boldsymbol{\pi}$ for some matrix $\mathbf{A}$, in which case $\mathbf{Q}(\boldsymbol{\pi}) = \mathbf{A}$. Loglinear and logit response functions have the form $\mathbf{F}(\boldsymbol{\pi}) = \mathbf{K} \log (\mathbf{A}\boldsymbol{\pi})$ for certain matrices $\mathbf{K}$ and $\mathbf{A}$, where log transforms a vector to the corresponding vector of natural logarithms. In that case $\mathbf{Q}(\boldsymbol{\pi}) = \mathbf{K}\mathbf{D}^{-1}\mathbf{A}$, where $\mathbf{D}$ is a diagonal matrix with the elements of the vector $\mathbf{A}\boldsymbol{\pi}$ on the main diagonal.

By a multivariate version of the delta method discussed in Appendix C, the asymptotic variance of $\mathbf{F}(\mathbf{p})$, the sample version of $\mathbf{F}(\boldsymbol{\pi})$, is

$$\boldsymbol{\Sigma} = \mathbf{Q}(\boldsymbol{\pi})\mathbf{V}(\boldsymbol{\pi})\mathbf{Q}(\boldsymbol{\pi})'$$

For large n, $\boldsymbol{\Sigma}$ can be estimated by its sample value $\mathbf{S} = \mathbf{Q}(\mathbf{p})\mathbf{V}(\mathbf{p})\mathbf{Q}(\mathbf{p})'$. In what follows, we assume that the $u \times u$ matrix $\mathbf{S}$ is nonsingular. For models in which $u = s(c-1)$, such as the cumulative logit models, this requires that all $p_{j(i)} > 0$. It is common to add a constant (e.g., 0.5) to all n_{ij} to ensure this.

We consider a linear model

$$\mathbf{F}(\boldsymbol{\pi}) = \mathbf{X}\boldsymbol{\beta}$$

for the functions $\mathbf{F}$, where $\mathbf{X}$ is a $u \times v$ design matrix of known constants having rank v, and $\boldsymbol{\beta}$ is a $v \times 1$ vector of parameters. The WLS estimate of $\boldsymbol{\beta}$ is the vector that minimizes the quadratic form

$$(\mathbf{F}(\mathbf{p}) - \mathbf{X}\boldsymbol{\beta})'\mathbf{S}^{-1}(\mathbf{F}(\mathbf{p}) - \mathbf{X}\boldsymbol{\beta})$$

This estimate equals

$$\mathbf{b} = (\mathbf{X}'\mathbf{S}^{-1}\mathbf{X})^{-1}\mathbf{X}'\mathbf{S}^{-1}\mathbf{F}(\mathbf{p})$$

The estimated asymptotic covariance matrix of $\mathbf{b}$ is $(\mathbf{X}'\mathbf{S}^{-1}\mathbf{X})^{-1}$.

The goodness of fit of the model can be tested using the residual term

$$X^2 = (\mathbf{F}(\mathbf{p}) - \mathbf{X}\mathbf{b})'\mathbf{S}^{-1}(\mathbf{F}(\mathbf{p}) - \mathbf{X}\mathbf{b})$$
$$= \mathbf{F}(\mathbf{p})'\mathbf{S}^{-1}\mathbf{F}(\mathbf{p}) - \mathbf{b}'(\mathbf{X}'\mathbf{S}^{-1}\mathbf{X})\mathbf{b}$$

Under the null hypothesis (H_0: $\mathbf{F}(\boldsymbol{\pi}) - \mathbf{X}\boldsymbol{\beta} = 0$), X^2 has an asymptotic chi-squared distribution with df $= u - v$. The degrees of freedom equal the difference between the number of response functions and the number of parameters in the model.

A hypothesis about effects of explanatory variables can be expressed in the form H_0: $\mathbf{C}\boldsymbol{\beta} = 0$, where $\mathbf{C}$ is a $d \times v$ matrix of constants of rank d. This can be tested using

$$X^2 = \mathbf{b}'\mathbf{C}'[\mathbf{C}(\mathbf{X}'\mathbf{S}^{-1}\mathbf{X})^{-1}\mathbf{C}']^{-1}\mathbf{C}\mathbf{b}$$

which has an asymptotic chi-squared distribution with df $= d$ if H_0 is true. Note that the expression inside the brackets of X^2 is simply the estimated asymptotic covariance matrix of $\mathbf{Cb}$.

A.2. WLS FOR LOGIT MODELS

We shall first illustrate the WLS approach for the cumulative logit models of Chapter 7. Consider the row effects model (7.6) applied to the political ideology data of Table 7.2; that is,

$$L_{j(i)} = \alpha_j + \tau_i, \qquad i = 1, 2, 3, \qquad j = 1, 2$$

where $\sum \tau_i = 0$. The response vector $\mathbf{F}(\boldsymbol{\pi})$ consists of the six logits,

$$\mathbf{F}' = \mathbf{F}(\boldsymbol{\pi})' = (L_{1(1)}, L_{2(1)}, L_{1(2)}, L_{2(2)}, L_{1(3)}, L_{2(3)})$$

The model is $\mathbf{F} = \mathbf{X}\boldsymbol{\beta}$, with $\boldsymbol{\beta}' = (\alpha_1, \alpha_2, \tau_1, \tau_2)$ and

$$\mathbf{X} = \begin{pmatrix} 1 & 0 & 1 & 0 \\ 0 & 1 & 1 & 0 \\ 1 & 0 & 0 & 1 \\ 0 & 1 & 0 & 1 \\ 1 & 0 & -1 & -1 \\ 0 & 1 & -1 & -1 \end{pmatrix}$$

since $\tau_3 = -\tau_1 - \tau_2$.

Let $\mathbf{L}_i' = (L_{1(i)}, L_{2(i)})$ be the two logits formed in row i of the 3×3 table. Let $\boldsymbol{\pi}' = (\boldsymbol{\pi}_{(1)}', \boldsymbol{\pi}_{(2)}', \boldsymbol{\pi}_{(3)}')$, where $\boldsymbol{\pi}_{(i)}' = (\pi_{1(i)}, \pi_{2(i)}, \pi_{3(i)})$. We can express $\mathbf{L}_i$ in terms of $\boldsymbol{\pi}_{(i)}$ by

$$\mathbf{L}_i = \mathbf{K}^* \log (\mathbf{A}^* \boldsymbol{\pi}_{(i)})$$

where

$$\mathbf{A}^* = \begin{pmatrix} 1 & 0 & 0 \\ 0 & 1 & 1 \\ 1 & 1 & 0 \\ 0 & 0 & 1 \end{pmatrix} \quad \text{and} \quad \mathbf{K}^* = \begin{pmatrix} -1 & 1 & 0 & 0 \\ 0 & 0 & -1 & 1 \end{pmatrix}$$

Similarly, the complete model $\mathbf{F} = \mathbf{X}\boldsymbol{\beta}$ can be expressed as

$$\mathbf{K} \log (\mathbf{A}\boldsymbol{\pi}) = \mathbf{X}\boldsymbol{\beta}$$

where $\mathbf{K}$ and $\mathbf{A}$ are the Kronecker products $\mathbf{K} = \mathbf{I} \otimes \mathbf{K}^*$, $\mathbf{A} = \mathbf{I} \otimes \mathbf{A}^*$, and $\mathbf{I}$ is the 3×3 identity matrix. The Kronecker product $\mathbf{C} \otimes \mathbf{D}$ of a $r_1 \times c_1$ matrix $\mathbf{C}$ and a $r_2 \times c_2$ matrix $\mathbf{D}$ denotes the $r_1 r_2 \times c_1 c_2$ matrix

$$\mathbf{C} \otimes \mathbf{D} = \begin{pmatrix} c_{11}\mathbf{D} & c_{12}\mathbf{D} & \cdots \\ c_{21}\mathbf{D} & c_{22}\mathbf{D} & \cdots \\ \vdots & \vdots & \end{pmatrix}$$

To fit the model, we assume independent multinomial sampling and obtain the estimated covariance matrix $\mathbf{V}(\mathbf{p})$ of $\mathbf{p}$. The estimated asymptotic covariance matrix of the vector of sample logits $\mathbf{F}(\mathbf{p})$ is $\mathbf{S} = \mathbf{QVQ}'$, where $\mathbf{Q} = \mathbf{KD}^{-1}\mathbf{A}$ and $\mathbf{D}$ is the diagonal matrix with the elements of the vector $\mathbf{Ap}$ on the main diagonal. The WLS estimate of $\boldsymbol{\beta}$ is

$$\mathbf{b} = (\mathbf{X}'\mathbf{S}^{-1}\mathbf{X})^{-1}\mathbf{X}'\mathbf{S}^{-1}\mathbf{F}(\mathbf{p}) = (0.528, -1.315, -0.665, -0.280)'$$

The residual chi-squared for testing goodness of fit is $X^2 = 4.53$, with $\text{df} = 6 - 4 = 2$. The test of independence (H_0: $\tau_1 = \tau_2 = \tau_3 = 0$) can be expressed as $\mathbf{C\beta} = \mathbf{0}$, with

$$\mathbf{C} = \begin{pmatrix} 0 & 0 & 1 & 0 \\ 0 & 0 & 0 & 1 \end{pmatrix}$$

Its test statistic is $X^2 = 91.56$ based on df = 2. The results here are quite similar to those obtained with ML estimation. In Section 7.2 we obtained the parameter estimates $\hat{\boldsymbol{\beta}}' = (0.532, \ -1.325, \ -0.670, \ -0.282)$, the goodness-of-fit statistic $G^2 = 4.70$, and the statistic $G^2(I \mid R) = 100.96$ for the test of independence.

To apply the cumulative logit model $L_{j(i)} = \alpha_j + \beta(u_i - \bar{u})$ to the dumping severity data in Table 7.1, take $\mathbf{F} = \mathbf{X\beta}$ with $\mathbf{F}' = (L_{1(1)}, L_{2(1)}, L_{1(2)}, L_{2(2)}, L_{1(3)}, L_{2(3)}, L_{1(4)}, L_{2(4)})$ and $\boldsymbol{\beta}' = (\alpha_1, \alpha_2, \beta)$. The uniform association version of the model (for which $u_i = i$) has

$$\mathbf{X} = \begin{pmatrix} 1 & 0 & -1.5 \\ 0 & 1 & -1.5 \\ 1 & 0 & -0.5 \\ 0 & 1 & -0.5 \\ 1 & 0 & 0.5 \\ 0 & 1 & 0.5 \\ 1 & 0 & 1.5 \\ 0 & 1 & 1.5 \end{pmatrix}$$

Proceeding as shown for the row effects model, we obtain $\mathbf{b}' = (-0.323, -2.034, 0.222)$ for the WLS estimate of $\boldsymbol{\beta}$, and the residual X^2 is 4.56 based on df = 5. Here $\mathbf{C} = (0, 0, 1)$ and $X^2 = 6.38$ based on df = 1 for testing H_0: $\beta = 0$. These results are similar to the ones obtained in Section 7.1 for ML estimation. For further discussion of the use of WLS estimation for cumulative logit models, see Williams and Grizzle (1972).

A.3. WLS FOR LOGLINEAR MODELS

The WLS approach is also simple to use for fitting ordinal loglinear models. The ordinal models discussed in Chapter 5 can be characterized by their local odds ratios. For two-way tables let the response vector $\mathbf{F}(\boldsymbol{\pi})'$ be $(\log \theta_{11}, \log \theta_{12}, \ldots, \log \theta_{r-1,c-1})$. The uniform association model states that $\log \theta_{ij}$ equals a constant, so it can be expressed as $\mathbf{F} = \mathbf{X}\beta$ where $\mathbf{X}$ is a column vector of $(r-1)(c-1)$ "1" elements. Here $s = 1$, and the response vector has the form $\mathbf{F} = \mathbf{K} \log(\mathbf{A}\boldsymbol{\pi})$, where $\boldsymbol{\pi}' = (\pi_{11}, \ldots, \pi_{rc})$, $\mathbf{A}$ is an $rc \times rc$ identity matrix, and $\mathbf{K}$ is a $(r-1)(c-1) \times rc$ matrix such that each row consists entirely of zero entries except for two $+1$ elements and two -1

elements, since

$$\log \theta_{ij} = \log \pi_{ij} + \log \pi_{i+1,j+1} - \log \pi_{i,j+1} - \log \pi_{i+1,j}$$

Hence the estimated asymptotic covariance matrix $\mathbf{S}$ of the sample log odds ratios, the WLS estimate b, and goodness-of-fit statistics can be obtained as described in A.1.

For the loglinear row effects model (5.5) with integer scores, the local log odds ratios have the form $\log \theta_{ij} = \beta_i$, with $\beta_i = \tau_{i+1} - \tau_i$. Thus $\boldsymbol{\beta}' = (\beta_1, \ldots, \beta_{r-1})$ and $\mathbf{F} = \mathbf{X}\boldsymbol{\beta}$, where the first $c-1$ rows of $\mathbf{X}$ equal (1, 0, 0, ..., 0), the next $c-1$ rows equal (0, 1, 0, ..., 0), and so forth. Like the uniform association model, the general uniform interaction model [(5.19) with integer scores] can be expressed as a single parameter model by letting $\mathbf{F}$ be the vector of $(r-1)(c-1)(l-1)$ local interaction measures $\log \theta_{ijk}$.

A.4. WLS FOR OTHER MODELS

The uniform association model (8.5) for the global odds ratios $\{\theta''_{ij}\}$ can also be expressed as $\mathbf{F} = \mathbf{K} \log(\mathbf{A}\boldsymbol{\pi}) = \mathbf{X}\beta$, where $\boldsymbol{\pi}' = (\pi_{11}, \ldots, \pi_{rc})$, $\mathbf{F}' = (\log \theta''_{11}, \ldots, \log \theta''_{r-1,c-1})$ and $\mathbf{X}$ is a column vector of $(r-1)(c-1)$ "1" elements. Now $\mathbf{A}$ is a $4(r-1)(c-1) \times rc$ matrix of 1 and 0 elements. Each block of four rows forms the four sums of probabilities for which the cross-product ratio is calculated for a particular θ''_{ij}. The matrix $\mathbf{K}$ consists entirely of zero entries except for two $+1$ and two -1 elements in each row, placed in proper position to form $\log \theta''_{ij}$ when applied to $\log(\mathbf{A}\boldsymbol{\pi})$. This matrix can be expressed as $\mathbf{K} = (1, -1, -1, 1) \otimes \mathbf{I}$ for an $(r-1)(c-1)$ identity matrix, when the first and fourth rows of every block in $\mathbf{A}$ sum the probabilities in the corners of the table at which both variables are high and at which both variables are low.

The WLS method can also be used to fit mean or mean ridit response models. For example, the regression model (8.8) for means $\{M_i = \sum_j v_j \pi_{j(i)}\}$ can be expressed as $\mathbf{F} = \mathbf{X}\boldsymbol{\beta}$ with $\mathbf{F}' = (M_1, M_2, \ldots, M_r)$, $\boldsymbol{\beta}' = (\mu, \beta)$, and

$$\mathbf{X} = \begin{pmatrix} 1 & u_1 - \bar{u} \\ 1 & u_2 - \bar{u} \\ \vdots & \vdots \\ 1 & u_r - \bar{u} \end{pmatrix}$$

$\mathbf{F}$ has the linear form $\mathbf{F} = \mathbf{A}\boldsymbol{\pi}$, so $\boldsymbol{\Sigma} = \mathbf{A}\mathbf{V}(\boldsymbol{\pi})\mathbf{A}'$ is the exact covariance matrix of the vector of sample means, and the WLS estimate is obtained in the regular way using $\mathbf{S} = \mathbf{A}\mathbf{V}(\mathbf{p})\mathbf{A}'$.

For the proportional hazards model (8.9) the response has the form $\mathbf{F}(\boldsymbol{\pi}) = \log(-\log(\mathbf{A}\boldsymbol{\pi}))$. Each row in the matrix $\mathbf{A}$ consists entirely of zeroes, except for "1" elements placed to give the sum of conditional probabilities corresponding to the complement of the distribution function. The estimated asymptotic covariance matrix of the sample vector $\mathbf{F}(\mathbf{p})$ is $\mathbf{S} = \mathbf{QVQ}'$, where $\mathbf{Q} = \mathbf{D}_2^{-1}\mathbf{D}_1^{-1}\mathbf{A}$. Here $\mathbf{D}_1$ is the diagonal matrix with elements of the vector $\mathbf{Ap}$ on the main diagonal, and $\mathbf{D}_2$ is the diagonal matrix with elements of the vector $-\log(\mathbf{Ap})$ on the main diagonal. The WLS method can be used to fit the proportional hazards model by substituting this expression for $\mathbf{S}$ into the general formulas.

APPENDIX B

Maximum Likelihood Estimation

The first part of this appendix discusses maximum likelihood (ML) estimation for loglinear models. Three methods are presented for obtaining these estimates. In the second part a class of "generalized linear models" is discussed that contains loglinear models as a special case. This class of models can be easily fitted using a widely available computer package (GLIM), and its ML estimates exist and are unique under broad conditions. ML estimation of the cumulative logit models of Chapter 7 is briefly considered in the final part of this appendix.

B.1. ML FOR LOGLINEAR MODELS

Let $\mathbf{n} = (n_1, \ldots, n_u)$ and $\mathbf{m} = (m_1, \ldots, m_u)$ denote the observed counts and expected frequencies for the cells in the table. For simplicity we use a single index, though the table may be multidimensional. Consider the sampling model whereby the $\{n_i\}$ are independent Poisson random variables with $E(n_i) = m_i$. The likelihood function gives the probability of the observed sample as a function of the parameter $\mathbf{m}$, namely

$$L(\mathbf{m}) = \prod_i \frac{[\exp(-m_i)]m_i^{n_i}}{n_i!}$$

The ML estimate is the parameter value $\mathbf{m}$ that maximizes this function. Equivalently, it maximizes the log likelihood

$$\log L(\mathbf{m}) = c + \sum n_i \log m_i - \sum m_i \tag{B.1}$$

where the constant c is a function only of $\mathbf{n}$, not $\mathbf{m}$.

Loglinear models have the form

$$\log \mathbf{m} = \mathbf{X}\boldsymbol{\beta}$$

for a $u \times v$ design matrix of constants $\mathbf{X}$ and a vector of parameters $\boldsymbol{\beta}' = (\beta_1, \ldots, \beta_v)$ with $v \leq u$. Let x_{il} be the entry of $\mathbf{X}$ in row i and column l. Then $m_i = \exp\left(\sum_l x_{il}\beta_l\right)$ for a loglinear model. The ML estimate $\hat{\boldsymbol{\beta}}$ of $\boldsymbol{\beta}$ maximizes the log likelihood (B.1) expressed as a function of the model parameter $\boldsymbol{\beta}$, namely

$$\log L(\boldsymbol{\beta}) = c + \sum_i n_i\left(\sum_l x_{il}\beta_l\right) - \sum_i \exp\left(\sum_l x_{il}\beta_l\right)$$

It is obtained as the solution to the likelihood equations $\partial \log L(\boldsymbol{\beta})/\partial\beta_j = 0$, $j = 1, \ldots, v$. Now

$$\frac{\partial \log L(\boldsymbol{\beta})}{\partial \beta_j} = \sum_i n_i x_{ij} - \sum_i m_i x_{ij}, \qquad j = 1, \ldots, v$$

so the ML estimates $\{\hat{m}_i\}$ of $\{m_i\}$ satisfy

$$\mathbf{X}'\mathbf{n} = \mathbf{X}'\hat{\mathbf{m}} \tag{B.2}$$

This means that certain linear functions of expected cell frequencies are equated to sufficient statistics that are the same functions of observed cell counts. For qualitative terms in the loglinear model, (B.2) implies that certain marginal tables are the same for the $\{\hat{m}_i\}$ as for the $\{n_i\}$. For quantitative terms (B.2) implies equality for the $\{\hat{m}_i\}$ and $\{n_i\}$ in certain conditional and joint moments [e.g., see (5.3)].

The asymptotic covariance matrix of a ML estimate $\hat{\boldsymbol{\beta}}$ is the matrix whose inverse has element $-E(\partial^2 \log L(\boldsymbol{\beta})/\partial\beta_j\partial\beta_k)$ in row j and column k. For the loglinear model

$$\frac{\partial^2 \log L(\boldsymbol{\beta})}{\partial\beta_j\partial\beta_k} = -\sum_i x_{ij}\left(\frac{\partial m_i}{\partial \beta_k}\right) = -\sum_i x_{ij} m_i x_{ik}$$

and hence $-E(\partial^2 \log L(\boldsymbol{\beta})/\partial\beta_j\partial\beta_k) = \sum_i m_i x_{ij} x_{ik}$. Once we have calculated $\hat{\boldsymbol{\beta}}$, we can obtain $\hat{\mathbf{m}}$ through $\hat{\mathbf{m}} = \exp(\mathbf{X}\hat{\boldsymbol{\beta}})$ and we can estimate the covariance matrix of $\hat{\boldsymbol{\beta}}$ by

$$(\mathbf{X}'\hat{\mathbf{M}}\mathbf{X})^{-1}$$

where $\hat{\mathbf{M}}$ is the diagonal matrix with elements of $\hat{\mathbf{m}}$ on the main diagonal. Palmgren (1981) showed that the same asymptotic covariance matrix is obtained for multinomial sampling on a response variable (conditional on observed marginal totals for explanatory variables), for estimates of parameters involving that response.

For loglinear models the log likelihood function is concave, and parameter estimates necessarily exist and are unique and finite if all observed cell counts are positive. Unfortunately, for the ordinal loglinear models considered in Chapter 5, there are no closed-form expressions for $\hat{\boldsymbol{\beta}}$. We next discuss three methods of obtaining $\hat{\boldsymbol{\beta}}$ by iterative methods.

Newton-Raphson

The Newton-Raphson method is a procedure for finding the location at which a function is maximized. To use it, we must make an initial guess for that location. Then the function is approximated in a neighborhood of that guess by a second-degree polynomial, and the location of that polynomial's maximum value is the second guess. Next the function is approximated in the region of the second guess by another second-degree polynomial, and the third guess is the location of its maximum. In this manner a sequence of guesses is generated that converges to the location of the maximum if the function is suitably well behaved.

More specifically, suppose we want to find the value $\hat{\boldsymbol{\beta}}$ of $\boldsymbol{\beta}$ that maximizes a function $g(\boldsymbol{\beta})$. Let $\mathbf{q}' = (\partial g/\partial \beta_1, \partial g/\partial \beta_2, \ldots)$, let $\mathbf{H} = (h_{jk})$ with $h_{jk} = \partial^2 g/\partial \beta_j \beta_k$, and let $\mathbf{q}^{(t)}$ and $\mathbf{H}^{(t)}$ be those terms evaluated at $\boldsymbol{\beta}^{(t)}$, the tth iterate in the sequence of approximations for $\hat{\boldsymbol{\beta}}$. At the tth step of the iteration ($t = 0, 1, 2, \ldots$), $g(\boldsymbol{\beta})$ is approximated near $\boldsymbol{\beta}^{(t)}$ by the terms up to second order in its Taylor series expansion,

$$Q^{(t)}(\boldsymbol{\beta}) = g(\boldsymbol{\beta}^{(t)}) + \mathbf{q}^{(t)'}(\boldsymbol{\beta} - \boldsymbol{\beta}^{(t)}) + (\tfrac{1}{2})(\boldsymbol{\beta} - \boldsymbol{\beta}^{(t)})'\mathbf{H}^{(t)}(\boldsymbol{\beta} - \boldsymbol{\beta}^{(t)})$$

Solving $\partial Q^{(t)}/\partial \boldsymbol{\beta} = \mathbf{q}^{(t)} + \mathbf{H}^{(t)}(\boldsymbol{\beta} - \boldsymbol{\beta}^{(t)}) = \mathbf{0}$ for $\boldsymbol{\beta}$ yields the next approximation,

$$\boldsymbol{\beta}^{(t+1)} = \boldsymbol{\beta}^{(t)} - (\mathbf{H}^{(t)})^{-1}\mathbf{q}^{(t)} \tag{B.3}$$

assuming that $\mathbf{H}^{(t)}$ is nonsingular. See Bard (1974) for details.

For loglinear models we identify the log likelihood with $g(\boldsymbol{\beta})$ in the preceding formulation. Then

$$q_j = \frac{\partial \log L(\boldsymbol{\beta})}{\partial \beta_j} = \sum_i (n_i - m_i)x_{ij}, \qquad q_j^{(t)} = \sum_i (n_i - m_i^{(t)})x_{ij}$$

$$h_{jk} = \frac{\partial^2 \log L(\boldsymbol{\beta})}{\partial \beta_j \partial \beta_k} = -\sum_i m_i x_{ij} x_{ik}, \qquad h_{jk}^{(t)} = -\sum_i m_i^{(t)} x_{ij} x_{ik}$$

Here $\mathbf{m}^{(t)}$ (the tth approximation for $\hat{\mathbf{m}}$) is obtained from $\boldsymbol{\beta}^{(t)}$ through log $\mathbf{m}^{(t)} = \mathbf{X}\boldsymbol{\beta}^{(t)}$, and it is used to obtain the next value $\boldsymbol{\beta}^{(t+1)}$ through formula (B.3). This in turn is used to obtain $\mathbf{m}^{(t+1)}$, and so forth. Alternatively, $\boldsymbol{\beta}^{(t+1)}$

can be expressed as

$$\boldsymbol{\beta}^{(t+1)} = -(\mathbf{H}^{(t)})^{-1}\mathbf{r}^{(t)} \tag{B.4}$$

where

$$r_j^{(t)} = \sum_i m_i^{(t)} x_{ij} \left\{ \log m_i^{(t)} + \frac{(n_i - m_i^{(t)})}{m_i^{(t)}} \right\}$$

One can begin the iterative process by setting all $m_i^{(0)} = n_i$, or all $m_i^{(0)} = n_i + 0.5$ if any $n_i = 0$. Then (B.4) is used to obtain $\boldsymbol{\beta}^{(1)}$, and for $t \geq 1$ the iterations proceed as just described using (B.3). In the limit as t increases, $\mathbf{m}^{(t)}$ and $\boldsymbol{\beta}^{(t)}$ usually converge rapidly to the ML estimates $\hat{\mathbf{m}}$ and $\hat{\boldsymbol{\beta}}$. The $\mathbf{H}^{(t)}$ matrices then converge to the matrix $\hat{\mathbf{H}} = -\mathbf{X}'\hat{\mathbf{M}}\mathbf{X}$, so the estimated covariance matrix of $\hat{\boldsymbol{\beta}}$ is obtained as a by-product of this iterative scheme through $-\hat{\mathbf{H}}^{-1}$. For sufficiently large t the convergence of $\boldsymbol{\beta}^{(t+1)}$ to $\hat{\boldsymbol{\beta}}$ satisfies

$$|\beta_j^{(t+1)} - \hat{\beta}_j| \leq c\,|\beta_j^{(t)} - \hat{\beta}_j|^2 \quad \text{for some } c > 0$$

and is referred to as "second order." This is very fast, and it usually takes only three to five iterations to get a satisfactory fit for loglinear models.

Expression (B.4) can be expressed in the WLS form

$$\boldsymbol{\beta}^{(t+1)} = (\mathbf{X}'\mathbf{S}^{-1}\mathbf{X})^{-1}\mathbf{X}'\mathbf{S}^{-1}\mathbf{y} \tag{B.5}$$

where we identify $y_i = \log m_i^{(t)} + (n_i - m_i^{(t)})/m_i^{(t)}$ and $\mathbf{S}$ is the diagonal matrix of elements $1/m_i^{(t)}$. In other words, $\boldsymbol{\beta}^{(t+1)}$ is the WLS solution for the model

$$\mathbf{y} = \mathbf{X}\boldsymbol{\beta} + \boldsymbol{\varepsilon}$$

where the $\{\varepsilon_i\}$ are uncorrelated with variances $\{1/m_i^{(t)}\}$. Thus the ML estimate may be regarded as the limit of a sequence of WLS estimates. If we take $m_i^{(0)} = n_i$, then the first approximation for the ML estimate is simply the standard WLS estimate for the model

$$\log \mathbf{n} = \mathbf{X}\boldsymbol{\beta} + \boldsymbol{\varepsilon}$$

See Haberman (1978, pp. 64–77) for further details.

To illustrate the $\log \mathbf{m} = \mathbf{X}\boldsymbol{\beta}$ representation needed for fitting a loglinear model using the Newton-Raphson method, we consider the row effects model (5.5) applied to Table 5.3. Then $\log \mathbf{m} = \mathbf{X}\boldsymbol{\beta}$ with

$$\log \mathbf{m}' = (\log m_{11}, \log m_{12}, \log m_{13}, \log m_{21}, \ldots, \log m_{33})$$

$$\boldsymbol{\beta}' = (\mu, \lambda_1^X, \lambda_2^X, \lambda_1^Y, \lambda_2^Y, \tau_1, \tau_2)$$

and for integer column scores,

$$\mathbf{X} = \begin{pmatrix} 1 & 1 & 0 & 1 & 0 & -1 & 0 \\ 1 & 1 & 0 & 0 & 1 & 0 & 0 \\ 1 & 1 & 0 & -1 & -1 & 1 & 0 \\ 1 & 0 & 1 & 1 & 0 & 0 & -1 \\ 1 & 0 & 1 & 0 & 1 & 0 & 0 \\ 1 & 0 & 1 & -1 & -1 & 0 & 1 \\ 1 & -1 & -1 & 1 & 0 & 1 & 1 \\ 1 & -1 & -1 & 0 & 1 & 0 & 0 \\ 1 & -1 & -1 & -1 & -1 & -1 & -1 \end{pmatrix}$$

Here we use the fact that $\lambda_3^X = -\lambda_1^X - \lambda_2^X$, $\lambda_3^Y = -\lambda_1^Y - \lambda_2^Y$, and $\tau_3 = -\tau_1 - \tau_2$.

The computer packages GLIM and SPSS^X (LOGLINEAR routine), described in Appendixes D.1 and D.2, fit loglinear models using the Newton-Raphson method. Appendix D.4 contains a computer program that can be used with the SAS package to apply the Newton-Raphson method to ordinal loglinear models.

Unidimensional Newton-Raphson

Goodman (1979a) suggested another method for fitting ordinal loglinear models. At a particular step of a cycle one of the likelihood equations is treated as a function of a single model parameter so that it has the form $h(\beta_i) = 0$. The next approximation for the ML estimate $\hat{\beta}_i$ of that parameter is then obtained using Newton's method for approximating a root to $h(\beta_i) = 0$. This method is a unidimensional version of the Newton-Raphson routine. The sequence of approximations is

$$\beta_i^{(t+1)} = \beta_i^{(t)} - \frac{h(\beta_i^{(t)})}{h'(\beta_i^{(t)})}$$

where $\beta_i^{(t)}$ is the approximate solution at stage t.

To illustrate this method, we again consider the row effects model (5.5). For that model, again assuming independent Poisson sampling, (B.1) becomes

$$\begin{aligned} \log L(\boldsymbol{\beta}) &= c + \sum \sum n_{ij} \log m_{ij} - \sum \sum m_{ij} \\ &= c + \sum \sum n_{ij}[\mu + \lambda_i^X + \lambda_j^Y + \tau_i(v_j - \bar{v})] \\ &\quad - \sum \sum \exp\,[\mu + \lambda_i^X + \lambda_j^Y + \tau_i(v_j - \bar{v})] \end{aligned}$$

It follows that

$$\frac{\partial \log L(\boldsymbol{\beta})}{\partial \lambda_i^X} = n_{i+} - m_{i+}$$

$$\frac{\partial \log L(\boldsymbol{\beta})}{\partial \lambda_j^Y} = n_{+j} - m_{+j}$$

and

$$\frac{\partial \log L(\boldsymbol{\beta})}{\partial \tau_i} = \sum_j n_{ij}(v_j - \bar{v}) - \sum_j m_{ij}(v_j - \bar{v})$$

The ML estimates $\{\hat{m}_i\}$ must therefore satisfy the three sets of likelihood equations:

$$\begin{aligned} \hat{m}_{i+} - n_{i+} &= 0, \qquad i = 1, \ldots, r \\ \hat{m}_{+j} - n_{+j} &= 0, \qquad j = 1, \ldots, c \\ \sum_j \hat{m}_{ij} v_j - \sum_j n_{ij} v_j &= 0, \qquad i = 1, \ldots, r \end{aligned} \tag{B.6}$$

For convenience we express the row effects model in the multiplicative form

$$m_{ij} = \alpha_i \beta_j \gamma_i^{(v_j - \bar{v})}$$

where in (5.5) we identify $\mu + \lambda_i^X$ with $\log \alpha_i$, λ_j^Y with $\log \beta_j$, and τ_i with $\log \gamma_i$. Suppose that at a certain stage the approximate estimates are $m_{ij}^{(t)}$, $\alpha_i^{(t)}$, $\beta_j^{(t)}$, and $\gamma_i^{(t)}$. A new cycle proceeds as follows:

1. Treat the likelihood equation

$$n_{i+} - m_{i+} = 0 = n_{i+} - \alpha_i \sum_j \beta_j \gamma_i^{(v_j - \bar{v})}$$

as a function of α_i, say $h(\alpha_i)$. Then

$$\begin{aligned} h'(\alpha_i) &= -\sum_j \beta_j \gamma_i^{(v_j - \bar{v})} = -\left(\frac{1}{\alpha_i}\right) \sum_j \alpha_i \beta_j \gamma_i^{(v_j - \bar{v})} \\ &= -\left(\frac{1}{\alpha_i}\right) \sum_j m_{ij} = -\frac{m_{i+}}{\alpha_i} \end{aligned}$$

so by the Newton method the next approximation for $\hat{\alpha}_i$ is

$$\alpha_i^{(t+1)} = \alpha_i^{(t)} + \frac{\alpha_i^{(t)}(n_{i+} - m_{i+}^{(t)})}{m_{i+}^{(t)}} = \frac{\alpha_i^{(t)} n_{i+}}{m_{i+}^{(t)}}$$

$i = 1, \ldots, r$. Using the new approximation for $\hat{\alpha}_i$, recompute $m_{ij}^{(t)} = \alpha_i^{(t+1)} \beta_j^{(t)} (\gamma_i^{(t)})^{v_j - \bar{v}}$.

2. A similar argument, treating the second likelihood equation as a function of β_j, leads to

$$\beta_j^{(t+1)} = \frac{\beta_j^{(t)} n_{+j}}{m_{+j}^{(t)}}, \qquad j = 1, \ldots, c$$

Recompute $m_{ij}^{(t)} = \alpha_i^{(t+1)} \beta_j^{(t+1)} (\gamma_i^{(t)})^{v_j - \bar{v}}$.

3. Treating the third likelihood equation as a function of γ_i, we obtain

$$\gamma_i^{(t+1)} = \gamma_i^{(t)} \left\{ \frac{1 + \sum_j (v_j - \bar{v})(n_{ij} - m_{ij}^{(t)})}{\sum_j (v_j - \bar{v})^2 m_{ij}^{(t)}} \right\}$$

We then compute

$$m_{ij}^{(t+1)} = \alpha_i^{(t+1)} \beta_j^{(t+1)} (\gamma_i^{(t+1)})^{v_j - \bar{v}}$$

to obtain the approximations used for step 1 in the next cycle.

The iterations continue until the likelihood equations are satisfied sufficiently well. It is usual to begin the cycles by taking initial estimates of parameters equal to one so that the initial approximations for the $\{\hat{m}_{ij}\}$ all equal one. Once satisfactory convergence is obtained, the log transformation yields parameter estimates for the model expressed in additive form. The association parameter estimates can be shifted by a constant so that they satisfy constraints such as $\sum \hat{\tau}_i = 0$.

Consider now the linear-by-linear association model (5.1). In multiplicative form it is

$$m_{ij} = \alpha_i \beta_j \theta^{(u_i - \bar{u})(v_j - \bar{v})}$$

where $\log \theta = \beta$ in our notation for (5.1). The first two steps of the Newton method are the same as just given for the row effects model. For the third step we use

$$\theta^{(t+1)} = \theta^{(t)} \left\{ \frac{1 + \sum \sum (u_i - \bar{u})(v_j - \bar{v})(n_{ij} - m_{ij}^{(t)})}{\sum \sum (u_i - \bar{u})^2 (v_j - \bar{v})^2 m_{ij}^{(t)}} \right\}$$

which is based on the likelihood equation

$$\sum \sum (u_i - \bar{u})(v_j - \bar{v}) n_{ij} - \sum \sum (u_i - \bar{u})(v_j - \bar{v}) \hat{m}_{ij} = 0$$

The calculations in this procedure are very simple compared to those in the multidimensional Newton-Raphson method, since no matrix inversion is necessary. Disadvantages of this method are that convergence is slower and the covariance matrix of $\hat{\boldsymbol{\beta}}$ is not obtained as a by-product of the fitting procedure.

Iterative Scaling

The iterative proportional fitting (IPF) procedure introduced in Section 4.4, when modified slightly, can also be used to fit the ordinal loglinear models discussed in Chapter 5. The IPF method generates, by a scaling process, a sequence of cell counts $\{\mathbf{m}^{(t)}\}$ that satisfies successively each likelihood equation. This sequence converges to the ML estimate $\hat{\mathbf{m}}$ of $\mathbf{m}$. The generalization of this scaling method that can be applied to ordinal loglinear models is based on corollary 2 to theorem 1 of Darroch and Ratcliff (1972). For two-dimensional tables their result takes the following form:

Consider k sets of constraints, where the lth set has the form

$$\sum_i \sum_j a_{s(ij)}^{(l)} \hat{\pi}_{ij} = h_s^{(l)}, \qquad s = 1, \ldots, d(l)$$

where

$$\sum_{s=1}^{d(l)} a_{s(ij)}^{(l)} = 1, \quad \sum_{s=1}^{d(l)} h_s^{(l)} = 1, \quad \text{and} \quad a_{s(ij)}^{(l)} \geq 0, \quad h_s^{(l)} > 0$$

If there is a solution $\{\hat{\pi}_{ij}\}$ that satisfies the constraints, then there is a unique positive solution $\hat{\boldsymbol{\pi}}$ that satisfies them and is of the form

$$\hat{\pi}_{ij} = \pi_{ij}^{(0)} \prod_{l=1}^{k} \prod_{s=1}^{d(l)} [\lambda_s^{(l)}]^{a_{s(ij)}^{(l)}}$$

The solution $\hat{\boldsymbol{\pi}}$ is the limit of the sequence $\{\boldsymbol{\pi}^{(t)}\}$ defined by

$$\pi_{ij}^{(t+1)} = \pi_{ij}^{(t)} \prod_{s=1}^{d(t_k)} \left(\frac{h_s^{(t_k)}}{h_s^{(t_k,\, t)}} \right)^{a_{s(ij)}^{(t_k)}}, \qquad t \geq 0 \tag{B.7}$$

where t_k denotes the remainder after dividing t by k, and

$$h_s^{(l,t)} = \sum_i \sum_j a_{s(ij)}^{(l)} \pi_{ij}^{(t)}$$

We illustrate this method again using the row effects model (5.5). Take all initial values $\pi_{ij}^{(0)} = 1/rc$. The first set of likelihood equations in (B.6) can be expressed in terms of proportions as

$$\sum_i \sum_j a_{s(ij)}^{(1)} \hat{\pi}_{ij} = h_s^{(1)}, \quad s = 1, \ldots, r$$

where

$$\begin{aligned} a_{s(ij)}^{(1)} &= 1, \quad i = s, \qquad j = 1, \ldots, c \\ &= 0, \qquad \text{otherwise} \end{aligned}$$

and

$$h_s^{(1)} = p_{s+}$$

There are $d(1) = r$ equations in the first set of constraints.

The second set of likelihood equations in (B.6) can be expressed as

$$\sum_i \sum_j a_{s(ij)}^{(2)} \hat{\pi}_{ij} = h_s^{(2)}, \qquad s = 1, \ldots, c$$

where

$$a_{s(ij)}^{(2)} = 1, \qquad j = s, \qquad i = 1, \ldots, r$$
$$= 0 \qquad \text{otherwise}$$

and

$$h_s^{(2)} = p_{+s}$$

There are $d(2) = c$ equations in the second set of constraints.

The third set of likelihood equations in (B.6) can be expressed as

$$\sum_i \sum_j a_{s(ij)}^{(3)} \hat{\pi}_{ij} = h_s^{(3)}, \qquad s = 1, \ldots, r$$

where

$$a_{s(ij)}^{(3)} = v_j^*, \qquad i = s, \qquad j = 1, \ldots, c$$
$$= 0, \qquad \text{otherwise}$$

and

$$h_s^{(3)} = \sum_j p_{sj} v_j^*$$

Here the $\{v_j^*\}$ are a linear rescaling of the $\{v_j\}$ to the interval [0, 1]. In order that $\sum_s a_{s(ij)}^{(3)} = 1$ and $\sum_s h_s^{(3)} = 1$, we add the additional r equations to the third set of constraints;

$$\sum_i \sum_j a_{s(ij)}^{(3)} \hat{\pi}_{ij} = h_s^{(3)}, \qquad s = r + 1, \ldots, 2r$$

where

$$a_{s(ij)}^{(3)} = 1 - v_j^*, \qquad i = s - r, \qquad j = 1, \ldots, c$$
$$= 0, \qquad \text{otherwise}$$

and

$$h_s^{(3)} = \sum_j p_{sj}(1 - v_j^*)$$

There are $d(3) = 2r$ equations in the third set of constraints.

Substitution of these expressions into (B.7) gives the iterative scaling routine listed in Section 5.2 for the row effects model. There the expressions are given in the equivalent form for expected frequencies rather than cell probabilities. The analogous substitutions for the linear-by-linear association model (5.1) gives the iterative scaling routine listed in Section 5.1.

The calculations in this procedure are also relatively simple. However, the convergence tends to be very slow (first order), and the method does not produce the asymptotic covariance matrix of the estimates.

B.2. GENERALIZED LINEAR MODELS

Nelder and Wedderburn (1972) studied a class of models, called *generalized linear models*, that have the following properties:

1. The data are independent observations from a distribution in the exponential family (i.e., the probability density or mass function for an observation can be expressed as $a(\theta)b(y)\exp[t(y)\theta]$).
2. A linear model $\boldsymbol{\eta} = \mathbf{X}\boldsymbol{\beta}$ describes the data.
3. The expected value μ_i for the ith observation and the parameter η_i for the linear model are linked by $\eta_i = \phi(\mu_i)$ for some "link" function ϕ, $i = 1, \ldots, n$.

The Poisson sampling model for the cell count n_i has expected value $\mu_i = m_i$ and has probability function in the exponential family with $t(n_i) = n_i$ and $\theta_i = \log m_i$. Loglinear models have the form $\boldsymbol{\eta} = \mathbf{X}\boldsymbol{\beta}$ with $\boldsymbol{\eta}' = (\log m_1, \ldots)$. Hence loglinear models are special cases of generalized linear models in which the link function ϕ is the log function so that $\eta_i = \theta_i = \log m_i$.

Generalized linear models have several desirable properties. Maximum likelihood estimates for them exist and are unique under quite general conditions, due to powerful results for exponential families (see Andersen 1974). Nelder and Wedderburn (1972) showed how to fit generalized linear models using the Newton-Raphson method. The computer package GLIM is designed to do this. It is very simple to fit the ordinal loglinear models of Chapter 5 using that package, as shown in Appendix D.1. McCullagh and Nelder (1983) have recently written a book about generalized linear models. Jørgensen (1983) has proposed an even broader class of models that permits, for example, correlated errors and nonlinearity.

B.3. ML FOR LOGIT MODELS

McCullagh (1980) showed how to use the Newton-Raphson method for ML estimation of a class of models that includes the cumulative logit models of

Chapter 7 and the log-log models of Section 8.4. We now outline his approach for the special case of ML estimation of cumulative logit models. As in Appendix A, we assume independent multinomial samples of sizes $(n_1, \ldots, n_s)$ at s combinations of levels of explanatory variables, for a response variable having c levels.

We observed in Appendix A.1 that the logit models can be expressed in the form

$$\mathbf{F} = \mathbf{X}\boldsymbol{\beta}$$

$\mathbf{F}$ is the vector of the $s(c-1)$ true logits, $\mathbf{X}$ is a $s(c-1) \times v$ design matrix, and $\boldsymbol{\beta}' = (\beta_1, \ldots, \beta_v)$. The first $c-1$ of the β_i are parameters (denoted by α's in the models in Chapter 7) related to average logits. The jth logit (7.1) at the ith level of the explanatory variables is $L_{j(i)} = \log[(1 - F_{j(i)})/F_{j(i)}]$, where $F_{j(i)} = \pi_{1(i)} + \cdots + \pi_{j(i)}$. We denote the sample cumulative probabilities by

$$Z_{j(i)} = \frac{(n_{i1} + \cdots + n_{ij})}{n_i}$$

Let

$$\mathbf{X} = \begin{pmatrix} \mathbf{X}^{(1)} \\ \vdots \\ \mathbf{X}^{(s)} \end{pmatrix}$$

where $\mathbf{X}^{(i)} = (x_{jk}^{(i)})$ represents the $c-1$ rows of the design matrix pertaining to the logits at level i of the explanatory variables. It follows from the appendix in McCullagh (1980) that the derivative of the log likelihood is

$$\begin{aligned} q_k &= \frac{\partial \log L}{\partial \beta_k} \\ &= \sum_{i=1}^{s} n_i \left[\sum_{j=1}^{c-1} \frac{(Z_{j(i)}F_{j+1(i)} - Z_{j+1(i)}F_{j(i)})}{F_{j(i)}(F_{j+1(i)} - F_{j(i)})a_{ijk}} \right], \qquad k = 1, \ldots, v \end{aligned}$$

where

$$a_{ijk} = F_{j(i)}(1 - F_{j(i)})x_{jk}^{(i)} - F_{j(i)}(1 - F_{j+1(i)})x_{j+1,k}^{(i)}$$

The expected value of the second derivative is

$$\begin{aligned} h_{kl} &= E\left[\frac{\partial^2 \log L}{\partial \beta_k \beta_l}\right] \\ &= -\sum_{i=1}^{s} n_i \left\{ \sum_{j=1}^{c-1} \left[\frac{F_{j+1(i)}}{F_{j(i)}(F_{j+1(i)} - F_{j(i)})} \right] a_{ijk}\, a_{ijl} \right\} \end{aligned}$$

The Newton-Raphson method can be applied as described in Appendix

B.1. The sequence of estimates is

$$\boldsymbol{\beta}^{(t+1)} = \boldsymbol{\beta}^{(t)} - (\mathbf{H}^{(t)})^{-1}\mathbf{q}^{(t)}$$

where $\mathbf{q}^{(t)}$ and $\mathbf{H}^{(t)}$ denote $\mathbf{q}' = (q_1, \ldots, q_v)$ and $\mathbf{H} = (h_{kl})$ evaluated at estimated cumulative probabilities $\{F_{j(i)}^{(t)}\}$ obtained at the tth iteration. Alternatively, $\boldsymbol{\beta}^{(t+1)}$ can be expressed as

$$\boldsymbol{\beta}^{(t+1)} = -(\mathbf{H}^{(t)})^{-1}\mathbf{r}^{(t)}$$

where

$$\begin{aligned} r_k = \sum_{i=1}^{s} n_i \Bigg\{ \sum_{j=1}^{c-1} & [F_{j+1(i)}/F_{j(i)}(F_{j+1(i)} - F_{j(i)})]a_{ijk} \\ & \times [F_{j(i)}(1 - F_{j(i)}) \log (L_{j(i)}) - F_{j(i)}(1 - F_{j+1(i)}) \log (L_{j+1(i)}) \\ & - Z_{j(i)} + F_{j(i)}Z_{j+1(i)}/F_{j+1(i)}] \Bigg\} \end{aligned}$$

The initial estimates of $\beta_1, \ldots, \beta_{c-1}$ should be monotone decreasing. The log-likelihood function is concave, and good convergence usually occurs within three to five iterations.

APPENDIX C

Delta Method

In Section 10.3 the delta method was used to obtain the asymptotic distributions of measures of association. In this appendix we give the theoretical basis for that method, and we illustrate the derivation of the asymptotic variance for one of the ordinal measures of association (gamma).

C.1. ASYMPTOTIC NORMALITY AND THE DELTA METHOD

Under random sampling conditions many statistics have approximately normal distributions for large-sample sizes. For example, suppose that $\mathbf{X}_1$, $\mathbf{X}_2$, ... are independent, each having a k-dimensional distribution with mean $\boldsymbol{\mu}$ and covariance matrix $\boldsymbol{\Sigma}$. Suppose that $\bar{\mathbf{X}}$ is the vector of sample means $\bar{\mathbf{X}} = \sum \mathbf{X}_i/n$. Then by the multivariate version of the Central Limit Theorem (see Rao 1973, p. 128),

$$\sqrt{n}(\bar{\mathbf{X}} - \boldsymbol{\mu}) \xrightarrow{d} N(\mathbf{0}, \boldsymbol{\Sigma})$$

In other words, for large n, $\bar{\mathbf{X}}$ has approximately a multivariate normal distribution with mean $\boldsymbol{\mu}$ and covariance matrix $\boldsymbol{\Sigma}/n$.

The following very powerful and useful result of statistical limit theory states that, under weak conditions, functions of asymptotically normal statistics are themselves asymptotically normally distributed.

Suppose that the vector $\mathbf{T}_n' = (T_{n1}, \ldots, T_{nk})$ is asymptotically multivariate normal with mean $\boldsymbol{\theta}' = (\theta_1, \ldots, \theta_k)$ and covariance matrix $\boldsymbol{\Sigma}/n$, where $\boldsymbol{\Sigma} = (\sigma_{ij})$. Suppose the function $g(t_1, \ldots, t_k)$ has a nonzero differential at $\boldsymbol{\theta}$. Then

$$\sqrt{n}[g(\mathbf{T}_n) - g(\boldsymbol{\theta})] \xrightarrow{d} N(\mathbf{0}, \mathbf{d}'\boldsymbol{\Sigma}\mathbf{d}) \qquad \text{(C.1)}$$

where $\mathbf{d}' = (d_1, \ldots, d_k)$ with $d_i = (\partial g/\partial t_i)\big|_{\mathbf{t}=\boldsymbol{\theta}}$.

Thus $g(\mathbf{T}_n)$ is asymptotically normal with mean $g(\boldsymbol{\theta})$ and variance $\mathbf{d}'\boldsymbol{\Sigma}\mathbf{d}/n$. The proof of this result follows from the expansion

$$g(\mathbf{T}_n) - g(\boldsymbol{\theta}) = (\mathbf{T}_n - \boldsymbol{\theta})'\mathbf{d} + \varepsilon_n \| \mathbf{T}_n - \boldsymbol{\theta} \|$$

where $\varepsilon_n \xrightarrow{p} 0$ as $n \to \infty$. This expansion indicates that $g(\mathbf{T}_n) - g(\boldsymbol{\theta})$ behaves like a linear function of the approximately normal variate $(\mathbf{T}_n - \boldsymbol{\theta})$ for large n. See Rao (1973, p. 387) and Bishop et al. (1975, Sec. 14.6) for details. The method of using differentials to obtain the asymptotic variance of $g(\mathbf{T}_n)$ is called the *delta method.*

C.2. DELTA METHOD FOR MEASURES OF ASSOCIATION

The sample proportion p_{ij} for a cell in a cross-classification table can be expressed as $p_{ij} = \sum_l X_l/n$, where

$$\begin{aligned} X_l &= 1 \qquad \text{if } l\text{th observation in cell } (i, j) \\ &= 0, \qquad \text{otherwise} \end{aligned}$$

Thus the sample proportions $\mathbf{p}' = (p_{11}, \ldots, p_{rc})$ in a $r \times c$ table are sample means, and for random sampling they are asymptotically normally distributed. Since measures of association are functions of sample proportions, it follows from (C.1) that they typically have large-sample normal distributions. This application of the delta method was formulated by Goodman and Kruskal (1972) in the following form:

Let $\zeta = \zeta(\boldsymbol{\pi})$ denote a measure for a $r \times c$ cross-classification table that can be written as $z = v/\delta$, where v and δ are functions of the cell proportions $\{\pi_{ij}\}$. Let

$$v'_{ij} = \frac{\partial v}{\partial \pi_{ij}}, \qquad \delta'_{ij} = \frac{\partial \delta}{\partial \pi_{ij}}, \qquad \phi_{ij} = \delta v'_{ij} - v\delta'_{ij}$$

$i = 1, \ldots, r$ and $j = 1, \ldots, c$. Let $\hat{\zeta} = \zeta(\mathbf{p})$ denote the sample value of ζ for full multinomial sampling. Then

$$\sqrt{n}(\hat{\zeta} - \zeta) \to N(0, \sigma^2), \qquad \text{where}$$

$$\sigma^2 = \frac{\sum\sum \pi_{ij}\phi_{ij}^2 - (\sum\sum \pi_{ij}\phi_{ij})^2}{\delta^4} \tag{C.2}$$

The proof of this result follows from (C.1), letting $\mathbf{T}_n$ be the vector of sample proportions $\mathbf{p}$, $\boldsymbol{\theta}$ the vector of population proportions $\boldsymbol{\pi}$, and g the measure of association ζ. The vector $\mathbf{p}$ is asymptotically normal with mean

$\boldsymbol{\pi}$ and covariance matrix $\boldsymbol{\Sigma}/n$. The covariance of p_{ij} and p_{kl} is $\sigma_{ij,kl}/n$ with $\sigma_{ij,kl} = -\pi_{ij}\pi_{kl}$, and the variance of p_{ij} is $\sigma_{ij,ij}/n$ with $\sigma_{ij,ij} = \pi_{ij}(1-\pi_{ij})$. Now let $\mathbf{d}' = (d_{11}, \ldots, d_{rc})$ with $d_{ij} = \partial\zeta/\partial\pi_{ij}$. Then

$$d_{ij} = \frac{\partial(v/\delta)}{\partial\pi_{ij}} = \frac{(\delta v'_{ij} - v\delta'_{ij})}{\delta^2}$$

The asymptotic variance of $\sqrt{n}(\hat{\zeta} - \zeta)$ is $\mathbf{d}'\boldsymbol{\Sigma}\mathbf{d}$, which simplifies to expression (C.2).

In practice, it is necessary to estimate σ^2 by $\hat{\sigma}^2$, where $\hat{\sigma}^2$ is formula (C.2) with π_{ij} replaced by p_{ij} in all terms. This estimate is useful in large-sample inference. For example, $\hat{\zeta} \pm z_{p/2}\,\hat{\sigma}/\sqrt{n}$ is an approximate 100(1 − p) percent confidence interval for ζ. To test $H_0: \zeta = 0$, an obvious test statistic is $z = \sqrt{n}\hat{\zeta}/\hat{\sigma}$, which has an asymptotic standard normal null distribution.

C.3. ASYMPTOTIC VARIANCE OF GAMMA

We illustrate formula (C.2) by deriving it for the ordinal measure, gamma. Now $\gamma = v/\delta$ with $v = \Pi_c - \Pi_d$ and $\delta = \Pi_c + \Pi_d$, where

$$\Pi_c = 2\sum_{i<j}\sum_{k<l}\pi_{ij}\pi_{kl} \quad \text{and} \quad \Pi_d = 2\sum_{i<j}\sum_{k>l}\pi_{ij}\pi_{kl}$$

Let

$$\pi_{ij}^{(c)} = \sum_{a>i}\sum_{b>j}\pi_{ab} + \sum_{a<i}\sum_{b<j}\pi_{ab} \quad \text{and} \quad \pi_{ij}^{(d)} = \sum_{a>i}\sum_{b<j}\pi_{ab} + \sum_{a<i}\sum_{b>j}\pi_{ab}$$

The proportion π_{ij} occurs in Π_c in such a way that it is multiplied twice by every proportion in the sum $\pi_{ij}^{(c)}$. Hence $\partial\Pi_c/\partial\pi_{ij} = 2\pi_{ij}^{(c)}$, and similarly $\partial\Pi_d/\partial\pi_{ij} = 2\pi_{ij}^{(d)}$. It follows that

$$v'_{ij} = 2(\pi_{ij}^{(c)} - \pi_{ij}^{(d)}), \qquad \delta'_{ij} = 2(\pi_{ij}^{(c)} + \pi_{ij}^{(d)})$$

and

$$\phi_{ij} = \delta v'_{ij} - v\delta'_{ij} = 4(\Pi_d\pi_{ij}^{(c)} - \Pi_c\pi_{ij}^{(d)})$$

Finally,

$$\begin{aligned}\sum\sum \pi_{ij}\phi_{ij} &= 4\Pi_d\sum\sum\pi_{ij}\pi_{ij}^{(c)} - 4\Pi_c\sum\sum\pi_{ij}\pi_{ij}^{(d)} \\ &= 4\Pi_d\Pi_c - 4\Pi_c\Pi_d = 0\end{aligned}$$

In summary, $\sqrt{n}(\hat{\gamma} - \gamma) \xrightarrow{d} N(0, \sigma^2)$, with

$$\begin{aligned}\sigma^2 &= \frac{\sum\sum\pi_{ij}\phi_{ij}^2}{\delta^4} \\ &= \left[\frac{16}{(\Pi_c + \Pi_d)^4}\right]\sum\sum\pi_{ij}(\Pi_d\pi_{ij}^{(c)} - \Pi_c\pi_{ij}^{(d)})^2\end{aligned}$$

The expressions given for ϕ_{ij} in Section 10.3 for other measures of association are obtained in a similar manner. For further discussions of the methodology described in this appendix, see Goodman and Kruskal (1963, 1972), where results are also presented for independent multinomial sampling.

The reader should not conclude from this appendix that *all* functions of cell proportions are asymptotically normal. For instance, consider the measure $\phi^2 = \sum \sum (\pi_{ij} - \pi_{i+}\pi_{+j})^2/\pi_{i+}\pi_{+j}$. If the variables are independent, the result given in C.1 does not give an informative asymptotic distribution because the differential vector is identically zero. In that case the appropriate scaling factor is n (rather than $\sqrt{n}$), and $n(\hat{\phi}^2 - \phi^2) = n\hat{\phi}^2 = X^2$ has an asymptotic chi-squared (rather than normal) distribution.

APPENDIX D

Computer Packages for Cross-Classification Tables

This appendix is a guide to some computer packages that can be used for certain analyses of ordinal categorical data. It is with some misgivings that we include this section, since existing computer packages are continually updated and since new ones are regularly introduced. However, we have not placed much emphasis on computational details in this book, and we encourage readers to determine which analyses can be done at their institutions with existing software. The sample listings given here may help readers who have access to a certain package and who wish to see examples of input necessary for fitting ordinal models.

D.1. GLIM

The interactive computer package GLIM (Generalized Linear Interactive Modelling, see Baker and Nelder 1978) is a low-cost package sponsored by the Royal Statistical Society. It is designed for fitting generalized linear models (see Appendix B.2). This package can easily be applied to the fitting of loglinear models, under the assumption of independent Poisson sampling for the cell counts. GLIM automatically selects the log function for the "link" (since the log of the mean is the natural parameter in the exponential family representation) and uses log m_{ij} as the response in the linear model. The logit models discussed in Chapter 6 for dichotomous response variables can be fitted in GLIM by specifying a binomial sampling distribution, for which log $[\pi/(1 - \pi)]$ is the natural parameter. The proportional hazards model discussed in Section 8.4 can be fitted in GLIM by specifying the log-log link function.

We illustrate the use of GLIM for fitting the loglinear row effects model

Table D.1. GLIM Used for Fitting Loglinear Row Effects Model to Table 5.3

```
$UNITS 9
$FACTOR  PARTY 3  IDEOL 3
$DATA  COUNT  PARTY  IDEOL  V
$READ
             100    1    3    1
             156    1    2    0
             143    1    1   -1
             141    2    3    1
             210    2    2    0
             119    2    1   -1
             127    3    3    1
              72    3    2    0
              15    3    1   -1
$YVAR  COUNT
$ERROR  POISSON
$FIT   PARTY + IDEOL + PARTY.V  $
```

(5.5) to the political ideology data of Table 5.3. Table D.1 contains input that is sufficient for fitting the model. PARTY denotes party affiliation, IDEOL denotes political ideology, and V denotes the variable represented by the scores $v_j - \bar{v}$ that appear in the association term of the model. Here, we take $v_j = j$. The first three statements in Table D.1 indicate that there are nine observations (nine cell counts), that PARTY and IDEOL are qualitative variables each having three levels, and that each piece of data is a cell count followed by the levels of PARTY, IDEOL, and V for that count. After specifying that the cell counts are to be treated as Poisson variables, we fit the row effects model.

After three cycles of the Newton-Raphson iterative method, the value of $G^2 = 2.815$ (referred to as the "deviance" by GLIM) is printed. The independence model can be fitted by the statement

$FIT PARTY + IDEOL $

Among other options one can request parameter estimates, estimated expected values, and standardized residuals through the statement

$DISPLAY E R

The parameter estimates in GLIM are scaled so that $\hat{\lambda}_r^X = \hat{\lambda}_c^Y = \hat{\tau}_r = 0$. For the political ideology data, we get $\hat{\tau}_1 = -1.213$, $\hat{\tau}_2 = -0.943$, and $\hat{\tau}_3 = 0$. We can subtract the mean of these values from each $\hat{\tau}_i$ to obtain the values reported in Section 5.2, which sum to zero.

GLIM can also be used to fit the log-multiplicative row and column effects model discussed in Section 8.1, even though it is not loglinear. When either the $\{\mu_i\}$ or $\{v_j\}$ are fixed, then model (8.1) is loglinear. Hence, if we set all $v_j = j$, we can estimate the $\{\mu_i\}$ by fitting the loglinear row effects model. We then treat the $\{\hat{\mu}_i\}$ as fixed and estimate the $\{v_j\}$ as in a column effects model. We take $\{\hat{v}_j\}$ as our new (fixed) column scores and re-fit the row effects model to get the next estimate of the $\{\mu_i\}$. We continue in this manner until satisfactory convergence is achieved. Usually only a few cycles are required. The df value reported by GLIM will be incorrect, since at each stage it is based on treating one set of parameter scores as fixed. For the two-dimensional table, the correct $\text{df} = (r-2)(c-2)$. See Goodman (1979a) for a similar way of fitting this model using the unidimensional Newton-Raphson method.

D.2. SPSS^X

The computer package SPSS^X is a statistical package sponsored by SPSS Inc. (Statistical Package for the Social Sciences, see Nie et al. 1975). The LOGLINEAR program in this package gives ML fitting of loglinear models and the corresponding logit models using the Newton-Raphson iterative method. Optional output includes estimated expected frequencies, standardized and adjusted residuals, and model parameter estimates and their correlations. This program can be easily used to fit the standard loglinear and logit models described in Chapters 3, 4, and 6. It can also be used to fit the ordinal loglinear models of Chapter 5 and the equivalent logit models for adjacent categories (see Complement 7.8). Also zero weights can be attached to certain cells in order to fit models having structural zeros, such as the quasi-independence model.

We illustrate the use of SPSS^X (LOGLINEAR) for fitting the loglinear row effects model (5.5) to the political ideology data of Table 5.3. Table D.2 gives sufficient program statements. Each piece of data consists of the level of party affiliation (PARTY), the level of political ideology (IDEOL), and the corresponding cell count (FREQ). Models containing appropriate association terms for ordinal variables are specified by requesting orthogonal polynomial contrasts, as in Haberman (1974b). Here IDEOL is ordinal, as recognized by the statement

```
CONTRAST(IDEOL) = POLYNOMIAL /
```

The DESIGN statement specifies the form of the model. For instance,

```
DESIGN = IDEOL PARTY /
```

specifies the independence model, whereas

DESIGN = IDEOL PARTY IDEOL BY PARTY /

specifies the saturated model for two variables. Table D.2 has the statement

DESIGN = IDEOL PARTY PARTY BY IDEOL(1) /

which gives an association term with linear scores for ideology (i.e., the row effects model).

The ordinal scores used by SPSS^X are the coefficients of orthogonal polynomials. For instance, for the three categories of IDEOL in Table D.2, it uses $(-1/\sqrt{2}, 0, 1/\sqrt{2})$. Hence, if parameter estimates are requested for these data, they will be a factor of $\sqrt{2}$ larger than those reported in Section 5.2, which were based on scores $(-1, 0, 1)$. SPSS^X gives the option of specifying scores that are not equal interval. For instance, the phrase CONTRAST(IDEOL) = POLYNOMIAL(1, 2, 4)/ gives $v_3 - v_2 = 2(v_2 - v_1)$.

The linear-by-linear association model (5.1) could be fitted to the dumping severity data of Table 5.1 by the LOGLINEAR statement

```
LOGLINEAR  OPER(1,4) DUMP(1,3) /
  CONTRAST(OPER) = POLYNOMIAL /
  CONTRAST(DUMP) = POLYNOMIAL /
  DESIGN  OPER DUMP  OPER(1) BY DUMP(1) /
```

Table D.2. SPSS^X (LOGLINEAR) Used for Fitting Loglinear Row Effects Model to Table 5.3

```
DATA LIST LIST /  IDEOL PARTY FREQ *
VALUE LABELS IDEOL 1 'LIBERAL' 2 'MODERATE' 3 'CONSER' /
      PARTY 1 'DEMO' 2 'INDEP' 3 'REPUB'
WEIGHT BY FREQ
LOGLINEAR IDEOL(1,3) PARTY(1,3) /
      CONTRAST(IDEOL) = POLYNOMIAL /
      DESIGN = IDEOL PARTY PARTY BY IDEOL(1) /
BEGIN DATA
1   1   143
2   1   156
3   1   100
1   2   119
2   2   210
3   2   141
1   3    15
2   3    72
3   3   127
END DATA
```

The orthogonal polynomial scores $(-1/\sqrt{2}, 0, 1/\sqrt{2})$ for dumping severity (DUMP) and $(-1.5/\sqrt{5}, -0.5/\sqrt{5}, 0.5/\sqrt{5}, 1.5/\sqrt{5})$ for operation (OPER) yield a parameter estimate of $\hat{\beta} = 0.514$ from this program; this figure is $\sqrt{2 \times 5} = 3.16$ times larger than the value of $\hat{\beta} = 0.163$ reported in Section 5.1 for integer scores. Through iterative application, SPSS^X (LOGLINEAR) can also be used to fit the log-multiplicative model (8.1), in the same way as described for GLIM in the previous section.

D.3. BMDP

BMDP (Biomedical Computer Programs, see Dixon 1981) is a general purpose statistical package that has separate parts for various analyses. The BMDP-4F program performs analyses for cross-classification tables. It can be used to compute several of the measures of association discussed in Chapters 9 and 10 (e.g., gamma, Kendall's tau-*b*, Somers' *d*, Pearson correlation), and it computes asymptotic standard errors and tests whether the measures equal zero. It is also very good for fitting the loglinear models for nominal variables discussed in Chapters 3 and 4. In its present form it is not capable of fitting the ordinal loglinear and logit models discussed in Chapters 5 and 7. Table D.3 contains input for fitting various loglinear models to the death penalty data of Table 3.1 and for printing two-way marginal tables, estimates of expected values, and model parameter estimates.

D.4. SAS

Like BMDP, SAS (Statistical Analysis System, see Ray 1982) is a general purpose statistical program. The LOGIST procedure in SAS can be used for ML estimation of the standard logit models described in Chapter 6, even if some explanatory variables are continuous. The FUNCAT procedure gives WLS solutions for models for categorical data. The responses in those models are specified by the user through a sequence of linear, logarithmic, and exponential functions. FUNCAT can be used to fit the mean response model (8.8) and the logit models of Chapter 6. For the logit model FUNCAT also provides the option of a ML fit.

Table D.4 contains input sufficient for fitting the logit model (6.3) to the dumping severity data in which the second and third categories are combined to form a single category. This model is discussed in Section 6.3, and it is equivalent to the loglinear uniform association model for the collapsed

Table D.3. BMDP-4F Used for Fitting Loglinear Models to Table 3.1

```
/PROBLEM        TITLE IS 'DEATH PENALTY'.
/INPUT          VARIABLES ARE 3.
                FORMAT IS '(8F4.0)'.
                TABLE IS 2,2,2.
/VARIABLE       NAMES ARE PENALTY, VICTIM, DEFENDANT.
/TABLE          INDICES ARE PENALTY, VICTIM, DEFENDANT.
                SYMBOLS ARE P,V,D.
/CATEGORY       CODES(1) ARE 1,2.
                NAMES(1) ARE YES, NO.
                CODES(2) ARE 1,2.
                NAMES(2) ARE WHITE, BLACK.
                CODES(3) ARE 1,2.
                NAMES(3) ARE WHITE, BLACK.
/PRINT          MARGINAL IS 2.
                EXPECTED.
                LAMBDA.
/FIT            ALL.
/FIT            MODEL  IS P,V,D.
/FIT            MODEL  IS VD,P.
/FIT            MODEL  IS VD,VP.
/FIT            MODEL  IS VD,VP,DP.
/FIT            MODEL  IS PVD.
/END
  19 132   0   9  11  52   6  97
```

Table D.4. SAS Used for Fitting a Logit Model and Mean Response Model to Dumping Severity Data

```
DATA DUMPING;
INPUT OPER DUMP COUNT;
CARDS;
1   1   61
1   2   28
1   3    7
2   1   68
2   2   23
2   3   13
3   1   58
3   2   40
3   3   12
4   1   53
4   2   38
4   3   16
PROC PRINT;
PROC FUNCAT; WEIGHT COUNT;
DIRECT OPER;
MODEL DUMP = OPER / FREQ PROB P COVB X;
RESPONSE  1  -1  LOG  0  1  1 / 1  0  0;
RESPONSE -1   0  1;
```

Note: The second and third categories of dumping are combined for the logit model.

4×2 table. Operation is identified as a quantitative variable by the "DIRECT" statement. The second response statement produces a fit of the mean response model (8.8) for the original 4×3 table, using integer scores.

The MATRIX procedure in SAS can be used to write simple programs for analyses using matrix algebra. Table D.5 gives a program that fits loglinear models using the Newton-Raphson method discussed in Appendix B.1. The user supplies the observed cell counts following the CARDS statement and specifies the model by supplying a design matrix X. The data in Table D.5 are the political ideology data of Table 5.3. The design matrix is the 9×7 matrix given in Appendix B.1 for fitting the loglinear row effects model. In Table D.5 the observed cell counts are used as the initial approximations to $\hat{\mathbf{m}}$, and five cycles of the Newton-Raphson method are then applied. This program can be easily refined so that the number of cycles depends on the quality of convergence and so that various statistics are also computed and printed.

Table D.5. Using PROC MATRIX in SAS to Fit Loglinear Row Effects Model to Table 5.3, by the Newton-Raphson Method

```
DATA FREQ;
INPUT F1;
CARDS;
100
156
143
141
210
119
127
72
15
PROC MATRIX  ERRMAX = 50 FUZZ;
FETCH N DATA  =  FREQ;
X = J(9,7);
X(1, )  =   1   1   0   1   0  -1   0 ;
X(2, )  =   1   1   0   0   1   0   0 ;
X(3, )  =   1   1   0  -1  -1   1   0 ;
X(4, )  =   1   0   1   1   0   0  -1 ;
X(5, )  =   1   0   1   0   1   0   0 ;
X(6, )  =   1   0   1  -1  -1   0   1 ;
X(7, )  =   1  -1  -1   1   0   1   1 ;
X(8, )  =   1  -1  -1   0   1   0   0 ;
X(9, )  =   1  -1  -1  -1  -1  -1  -1 ;
NN  =  DIAG(N);
H  =  X'*NN*X;
L  =  LOG(N);
R  =  X'*NN*L;
B  =  (INV(H))*R;
M  =  EXP(X*B);
NOTE SKIP = 3 INITIAL M ESTIMATE IS; PRINT M;
NOTE SKIP = 3 INITIAL BETA ESTIMATE IS; PRINT B;
DO I  =  1 TO 5;
MM  =  DIAG(M);
Q  =  X'*(N - M);
H  =  X'*MM*X;
B1  =  B  +  (INV(H))*Q;
M  =  EXP(X*B1);
B  =  B1;
NOTE SKIP  =  6  ESTIMATED EXPECTED FREQUENCIES ARE; PRINT M;
NOTE SKIP  =  3 ESTIMATED BETA IS ; PRINT B1;
GSQ  =  2#N'*(L - LOG(M));
NOTE SKIP  =  3  G-SQUARED IS ; PRINT GSQ;
END;
```

D.5. OTHER PACKAGES

The package MULTIQUAL (Bock and Yates 1973) is specifically designed for ML fitting of loglinear and logit models to categorical data. It is particularly useful for ordinal data, since it can fit the ordinal loglinear and logit models discussed in Chapters 5 and 7. Information about MULTIQUAL is available from International Educational Services, P. O. Box 536, Mooresville, IN 46158.

ANOAS is a FORTRAN program for ML estimation of ordinal loglinear models for two-way tables. It uses the unidimensional version of the Newton-Raphson method discussed in Appendix B.1. ANOAS can be used to fit the linear-by-linear association model (5.1), the row effects model (5.5), the row and column effects model of Complement 5.6, the log-multiplicative model (8.1), the quasi-uniform association and quasi-independence models discussed in Section 11.1, and many other models. To obtain a copy of ANOAS, contact

Dr. Clifford C. Clogg
Population Issues Research Office
21 Burrowes Building
Pennsylvania State University
University Park, PA 16802

A somewhat different version of ANOAS has been developed by Dr. Leo Goodman, Department of Statistics, University of Chicago. His program allows the fitting of some of the complex models discussed in his (1981a) article, for example, the $R + C + RC$ model alluded to in Complement 8.2.

GENCAT (Landis et al. 1976) is a program for WLS estimation of models for categorical data. It is similar in nature to the SAS-FUNCAT procedure, but it can be used to fit the ordinal loglinear and logit models described in Chapters 5 and 7. Semenya and Koch (1980) described a wide variety of models for ordinal data for which GENCAT can be used.

FREQ is a FORTRAN program contained in the text by Haberman (1979) that fits loglinear models by ML (through Newton-Raphson) and also reports adjusted residuals. To use it for ordinal loglinear models, the models must be expressed in a somewhat different manner than in this text. See p. 377 of Haberman's text for the row effects model and p. 385 for the linear-by-linear association model.

ECTA (Fay and Goodman 1975) is a specialized package for ML estimation of loglinear models by iterative proportional fitting. It can be used to fit the ordinal loglinear models of Chapter 5 by an iterative sequence of steps in which frequencies that have the odds ratio pattern satisfied by the model are used as "starting values" at each step. To fit the uniform association model, for instance, starting values are given for which all local odds ratios are the same number. For a 4×3 table, for example, the starting values can be

$$\begin{matrix} 1 & 1 & 1 \\ 1 & \theta & \theta^2 \\ 1 & \theta^2 & \theta^4 \\ 1 & \theta^3 & \theta^6 \end{matrix}$$

for some $\theta > 0$. These frequencies are scaled to fit the observed marginal distributions, by requesting ECTA to fit the independence model. By trial and error, a sequence of common odds ratio values is used until the one is found that minimizes G^2. See Duncan and McRae (1979) for details. ECTA can be easily used to fit the quasi-independence model by giving starting values of 0 on the main diagonal and 1 off the diagonal.

SPSS (Statistical Package for the Social Sciences, see Nie et al. 1975) is a general computing package like BMDP and SAS. It can be used to calculate several measures of association and to give significance tests for two-way tables using the CROSSTABS routine.

APPENDIX E

Chi-Squared Distribution Values for Various Right-Hand Tail Probabilities

	α						
df	0.250	0.100	0.050	0.025	0.010	0.005	0.001
1	1.32330	2.70554	3.84146	5.02389	6.63490	7.87944	10.828
2	2.77259	4.60517	5.99147	7.37776	9.21034	10.5966	13.816
3	4.10835	6.25139	7.81473	9.34840	11.3449	12.8381	16.266
4	5.38572	7.77944	9.48773	11.1433	13.2767	14.8602	18.467
5	6.62568	9.23635	11.0705	12.8325	15.0863	16.7496	20.515
6	7.84080	10.6446	12.5916	14.4494	16.8119	18.5476	22.458
7	9.03715	12.0170	14.0671	16.0128	18.4753	20.2777	24.322
8	10.2188	13.3616	15.5073	17.5346	20.0902	21.9550	26.125
9	11.3887	14.6837	16.9190	19.0228	21.6660	23.5893	27.877
10	12.5489	15.9871	18.3070	20.4831	23.2093	25.1882	29.588
11	13.7007	17.2750	19.6751	21.9200	24.7250	26.7569	31.264
12	14.8454	18.5494	21.0261	23.3367	26.2170	28.2995	32.909
13	15.9839	19.8119	22.3621	24.7356	27.6883	29.8194	34.528
14	17.1170	21.0642	23.6848	26.1190	29.1413	31.3193	36.123
15	18.2451	22.3072	24.9958	27.4884	30.5779	32.8013	37.697
16	19.3688	23.5418	26.2962	28.8454	31.9999	34.2672	39.252
17	20.4887	24.7690	27.5871	30.1910	33.4087	35.7185	40.790
18	21.6049	25.9894	28.8693	31.5264	34.8053	37.1564	42.312
19	22.7178	27.2036	30.1435	32.8523	36.1908	38.5822	43.820
20	23.8277	28.4120	31.4104	34.1696	37.5662	39.9968	45.315
21	24.9348	29.6151	32.6705	35.4789	38.9321	41.4010	46.797
22	26.0393	30.8133	33.9244	36.7807	40.2894	42.7956	48.268

df	α						
	0.250	0.100	0.050	0.025	0.010	0.005	0.001
23	27.1413	32.0069	35.1725	38.0757	41.6384	44.1813	49.728
24	28.2412	33.1963	36.4151	39.3641	42.9798	45.5585	51.179
25	29.3389	34.3816	37.6525	40.6465	44.3141	46.9278	52.620
26	30.4345	35.5631	38.8852	41.9232	45.6417	48.2899	54.052
27	31.5284	36.7412	40.1133	43.1944	46.9630	49.6449	55.476
28	32.6205	37.9159	41.3372	44.4607	48.2782	50.9933	56.892
29	33.7109	39.0875	42.5569	45.7222	49.5879	52.3356	58.302
30	34.7998	40.2560	43.7729	46.9792	50.8922	53.6720	59.703
40	45.6160	51.8050	55.7585	59.3417	63.6907	66.7659	73.402
50	56.3336	63.1671	67.5048	71.4204	76.1539	79.4900	86.661
60	66.9814	74.3970	79.0819	83.2976	88.3794	91.9517	99.607
70	77.5766	85.5271	90.5312	95.0231	100.425	104.215	112.317
80	88.1303	96.5782	101.879	106.629	112.329	116.321	124.839
90	98.6499	107.565	113.145	118.136	124.116	128.299	137.208
100	109.141	118.498	124.342	129.561	135.807	140.169	149.449

Source: *Public Program Analysis*, by R. N. Forthofer and R. G. Lehnen, © 1981 by Lifetime Learning Publications, Belmont, California 94002, a division of Wadsworth, Inc. Reprinted by permission of the publisher.

Bibliography

Agresti, A. 1976. "The Effect of Category Choice on Some Ordinal Measures of Association." *J. Amer. Statist. Assoc.* **71:** 49–55.

Agresti, A. 1977. "Considerations in Measuring Partial Association for Ordinal Categorical Data." *J. Amer. Statist. Assoc.* **72:** 37–45.

Agresti, A. 1980. "Generalized Odds Ratios for Ordinal Data." *Biometrics* **36:** 59–67.

Agresti, A. 1981a. "Measures of Nominal-Ordinal Association." *J. Amer. Statist. Assoc.* **76:** 524–529.

Agresti, A. 1981b. "A Hierarchical System of Interaction Measures for Multidimensional Contingency Tables." *J. Roy. Statist. Soc.* **B 43:** 293–301.

Agresti, A. 1983a. "A Survey of Strategies for Modeling Cross-Classifications Having Ordinal Variables." *J. Amer. Statist. Assoc.* **78:** 184–198.

Agresti, A. 1983b. "Testing Marginal Homogeneity for Ordinal Categorical Variables." *Biometrics* **39:** 505–510.

Agresti, A. 1983c. "A Simple Diagonals-Parameter Symmetry and Quasisymmetry Model." *Statist. Probability Letters* **1:** 313–316.

Agresti, A., and B. Agresti. 1979. *Statistical Methods for the Social Sciences.* San Francisco: Dellen.

Agresti, A., and A. Kezouh. 1983. "Association Models for Multidimensional Cross-Classifications of Ordinal Variables." *Commun. Statist.* **A 12:** 1261–1276.

Agresti, A., and D. Wackerly. 1977. "Some Exact Conditional Tests of Independence for $R \times C$ Cross-Classification Tables." *Psychometrika* **42:** 111–125.

Aitchison, J., and S. D. Silvey. 1957. "The Generalization of Probit Analysis to the Case of Multiple Responses." *Biometrika* **44:** 131–140.

Andersen, A. H. 1974. "Multidimensional Contingency Tables." *Scan. J. Statist.* **1:** 115–127.

Andersen, E. B. 1980. *Discrete Statistical Models with Social Science Application.* Amsterdam: North-Holland.

Anderson, J. A., 1984. "Regression and Ordered Categorical Variables." *J. Roy. Statist. Soc.* **B 46.**

Anderson, J. A., and P. R. Philips. 1981. "Regression, Discrimination, and Measurement Models for Ordered Categorical Variables." *J. Roy. Statist. Soc.* **C 30:** 22–31.

Anderson, R. J., and J. R. Landis. 1982. "CATANOVA for Multidimensional Contingency Tables: Ordinal-Scale Response." *Commun. Statist.* **A 11:** 257–270.

Anderson, S., A. Auquier, W. W. Hauck, D. Oakes, W. Vandaele, and H. I. Weisberg. 1980. *Statistical Methods for Comparative Studies.* New York: Wiley.

Andrich, D. 1978. "A Rating Formulation for Ordered Response Categories." *Psychometrika* **43:** 561–573.

Andrich, D. 1979. "A Model for Contingency Tables Having an Ordered Response Classification." *Biometrics* **35:** 403–415.

Anscombe, F. 1981. *Computing in Statistical Science through APL.* New York: Springer-Verlag.

Aranda-Ordaz, F. J. 1983. "An Extension of the Proportional Hazards Model for Grouped Data." *Biometrics* **39:** 109–117.

Armitage, P. 1955. "Tests for Linear Trends in Proportions and Frequencies." *Biometrics* **11:** 375–386.

Baker, R. J., and J. A. Nelder. 1978. *The GLIM System. Release 3. Generalised Linear Interactive Modelling Manual.* Oxford: N.A.G.

Bard, Y. 1974. *Nonlinear Parameter Estimates.* New York: Academic.

Bartholomew, D. J. 1959. "A Test of Homogeneity for Ordered Alternatives." *Biometrika* **46:** 36–48.

Bartholomew, D. J. 1980. "Factor Analysis for Categorical Data" (with discussion). *J. Roy. Statist. Soc.* **B 42:** 293–321.

Benedetti, J. K., and M. B. Brown. 1978. "Strategies for the Selection of Loglinear Models." *Biometrics* **34:** 680–686.

Benzécri, J. P. 1976. *L'Analyse des Données, Tome 2: Correspondences.* Paris: Dunod.

Berry, K. J., P. W. Mielke Jr., and R. B. Jacobsen. 1977. "A Large Sample Procedure for Testing Coefficients of Ordinal Association: Goodman and Kruskal's Gamma and Somers' d_{ba} and d_{ab}." *Educ. Psychol. Measurement* **37:** 791–794.

Bhapkar, V. P. 1966. "A Note on the Equivalence of Two Criteria for Hypotheses in Categorical Data." *J. Amer. Statist. Assoc.* **61:** 228–235.

Bhapkar, V. P. 1968. "On the Analysis of Contingency Tables with a Quantitative Response." *Biometrics* **24:** 329–338.

Bickel, P. J., E. A. Hammel, and J. W. O'Connell. 1975. "Sex Bias in Graduate Admissions: Data from Berkeley." *Science* **187:** 398–403.

Birch, M. W. 1963. "Maximum Likelihood in Three-Way Contingency Tables." *J. Roy. Statist. Soc.* **B 25:** 220–233.

Birch, M. W. 1965. "The Detection of Partial Association, II: The General Case." *J. Roy. Statist. Soc.* **B 27:** 111–124.

Bishop, Y. M. M., S. E. Fienberg, and P. W. Holland. 1975. *Discrete Multivariate Analysis.* Cambridge: MIT Press.

Blalock, H. M. 1974. "Beyond Ordinal Measurement: Weak Tests of Stronger Theories." Chapter 15 in *Measurement in the Social Sciences.* Ed. by H. M. Blalock, Chicago: Aldine.

Blalock, H. M. 1976. "Can We Find a Genuine Ordinal Slope Analogue?" *Sociological Methodology.* San Francisco: Jossey-Bass, pp. 195–229.

Blyth, C. R. 1972. "On Simpson's Paradox and the Sure-Thing Principle." *J. Amer. Statist. Assoc.* **67:** 364–366.

Bock, R. D. 1975. *Multivariate Statistical Methods in Behavioral Research.* New York: McGraw-Hill.

Bock, R. D., and L. V. Jones. 1968. *The Measurement and Prediction of Judgement and Choice.* San Francisco: Holden-Day.

Bock, R. D., and G. Yates. 1973. "MULTIQUAL: Log-Linear Analysis of Nominal or Ordinal Qualitative Data by the Method of Maximum Likelihood." Chicago: International Educational Services.

Bonacich, P., and D. Kirby. 1976. "Establishing a metric from assumptions about linearity." *J. Amer. Statist. Assoc.* **71:** 56–61.

Box, J. F. 1978. *R. A. Fisher, The Life of a Scientist.* New York: Wiley.

Boyle, R. P. 1970. "Path Analysis and Ordinal Data," *Amer. J. Sociol.* **75:** 461–480.

Bradley, R. A., S. K. Katti, and I. J. Coons. 1962. "Optimal Scaling for Ordered Categories." *Psychometrika* **27:** 355–374.

Breslow, N. 1982. "Covariance Adjustment of Relative-Risk Estimates in Matched Studies." *Biometrics* **38:** 661–672.

Bross, I. D. J. 1958. "How to Use Ridit Analysis." *Biometrics* **14:** 18–38.

Brown, M. B. 1976. "Screening Effects in Multidimensional Contingency Tables." *J. Roy. Statist. Soc.* **C 25:** 37–46.

Brown, M. B., and J. K. Benedetti. 1977. "Sampling Behavior of Tests of Correlation in Two-Way Contingency Tables." *J. Amer. Statist. Assoc.* **72:** 309–315.

Brunden, M. N. 1972. "The Analysis of Non-Independent 2×2 Tables from $2 \times C$ Tables Using Rank Sums." *Biometrics* **28:** 603–607.

Burridge, J. 1981. "A Note on Maximum Likelihood Estimation for Regression Models Using Grouped Data." *J. Roy. Statist. Soc.* **B 43:** 41–45.

Clayton, D. G. 1974. "Some Odds Ratio Statistics for the Analysis of Ordered Categorical Data." *Biometrika* **61:** 525–531.

Clayton, D. G. 1976. "An Odds Ratio Comparison for Ordered Categorical Data with Censored Observations." *Biometrika* **63:** 405–408.

Clogg, C. C. 1982a. "Using Association Models in Sociological Research: Some Examples." *Amer. J. Sociol.* **88:** 114–134.

Clogg, C. C. 1982b. "Some Models for the Analysis of Association in Multiway Cross-Classifications Having Ordered Categories." *J. Amer. Statist. Assoc.* **77:** 803–815.

Cochran, W. G. 1954. "Some Methods of Strengthening the Common χ^2 tests." *Biometrics* **10:** 417–451.

Cochran, W. G. 1968. "The Effectiveness of Adjustment by Subclassification in Removing Bias in Observational Studies." *Biometrics* **24:** 295–313.

Cohen, J. 1960. "A Coefficient of Agreement for Nominal Scales." *Educ. Psychol. Measurement.* **20:** 37–46.

Conover, W. J. 1974. "Some Reasons for Not Using the Yates Continuity Correction on 2×2 Tables" (with comments). *J. Amer. Statist. Assoc.* **69:** 374–382.

Cornfield, J. 1962. "Joint Dependence of Risk of Coronary Heart Disease on Serum Cholesterol and Systolic Blood Pressure: A Discriminant Function Analysis." *Federation Proc.* **21:** 58–61.

Cox, D. R. 1970. *The Analysis of Binary Data.* London: Chapman and Hall.

Cox, D. R. 1972. "Regression Models and Life Tables" (with discussion). *J. Roy. Statist. Soc.* **B 34:** 187–220.

Cox, D. R., and D. V. Hinkley. 1974. *Theoretical Statistics.* London: Chapman and Hall.

Cramér, H. 1946. *Mathematical Methods of Statistics.* Princeton: Princeton University Press.

Crittenden, K. C., and A. C. Montgomery. 1980. "A System of Paired Asymmetric Measures of Association for Use with Ordinal Dependent Variables." *Social Forces* **58:** 1178–1194.

Daganzo, C. 1979. *Multinomial Probit: The Theory and Its Application to Demand Forecasting.* New York: Academic.

Dale, J. R. 1982. "Global Cross-Ratio Models for Bivariate, Discrete, Ordered Response." Unpublished manuscript.

Daniels, H. E. 1944. "The Relation between Measures of Correlation in the Universe of Sample Permutations." *Biometrika* **33:** 129–135.

Darroch, J. N., and D. Ratcliff. 1972. "Generalized Iterative Scaling for Log-Linear Models." *Ann. Math. Statist.* **43:** 1470–1480.

Das Gupta, S., and M. D. Perlman. 1974. "Power of the Noncentral F-test: Effect of Additional Variates on Hotelling's T^2-Test." *J. Amer. Statist. Assoc.* **69:** 174–180.

Davis, J. A. 1967. "A Partial Coefficient for Goodman and Kruskal's Gamma." *J. Amer. Statist. Assoc.* **62:** 189–193.

Deming, W. E., and F. F. Stephan. 1940. "On a Least Squares Adjustment of a Sampled Frequency Table When the Expected Marginal Totals Are Known," *Ann. Math. Statist.* **11:** 427–444.

Dixon, W. J., ed. 1981. *BMDP Statistical Software 1981.* Berkeley: University of California Press.

Draper, N. R., and H. Smith. 1981. *Applied Regression Analysis.* 2d ed. New York: Wiley.

Duncan, O. D. 1979. "How Destination Depends on Origin in the Occupational Mobility Table." *Amer. J. Sociol.* **84:** 793–803.

Duncan, O. D., and J. A. McRae Jr., 1979. "Multiway Contingency Analysis with a Scaled Response or Factor." *Sociological Methodology.* San Francisco: Jossey-Bass, pp. 66–85.

Edwards, A. W. F. 1963. "The Measure of Association in a 2 × 2 Table" *J. Roy. Statist. Soc.* **A 126:** 109–114.

Efron, B., and C. Morris. 1977. "Stein's Paradox in Statistics." *Sci. Amer.* **236:** 119–127.

Everitt, B. S. 1977. *The Analysis of Contingency Tables.* London: Chapman and Hall.

Farewell, V. T. 1982. "A Note on Regression Analysis of Ordinal Data with Variability of Classification." *Biometrika* **69:** 533–538.

Fay, R., and L. A. Goodman. 1975. "ECTA Program: Description for Users." Chicago: Department of Statistics, University of Chicago.

Fienberg, S. 1980. *The Analysis of Cross-Classified Categorical Data.* 2d ed. Cambridge: MIT Press.

Fienberg, S. 1982. "Using Information on Ordering for Loglinear Model Analysis of Multidimensional Contingency Tables." *Proc. 11th International Biometrics Conference.*

Fienberg, S. E., and W. M. Mason. 1979. "Identification and Estimation of Age-Period-Cohort Models in the Analysis of Discrete Archival Data." *Sociological Methodology.* San Francisco: Jossey-Bass, pp. 1–67.

Finney, D. J., 1971. *Probit Analysis.* 3d ed. Cambridge: Cambridge University Press.

Fisher, R. A. 1934. *Statistical Methods for Research Workers.* 5th ed. (14th ed., 1970). Edinburgh: Oliver and Boyd.

Fix, E., J. L. Hodges, and E. L. Lehmann. 1959. "The Restricted Chi-Square Test." In *Probability and Statistics: The Harald Cramer Volume.* Ed. U. Grenander. New York: Wiley.

Fleiss, J. L. 1979. "Confidence Intervals for the Odds Ratio in Case-Control Studies: The State of the Art." *J. Chron. Dis.* **32:** 69–77.

Fleiss, J. L. 1981. *Statistical Methods for Rates and Proportions.* 2d ed. New York: Wiley-Interscience.

Fleiss, J. L. 1982. "A Simplification of the Classic Large-Sample Standard Error of a Function of Multinomial Proportions." *Amer. Statist.* **36:** 377–378.

Fleiss, J. L., J. Cohen, and B. S. Everitt. 1969. "Large-Sample Standard Errors of Kappa and Weighted Kappa." *Psychol. Bull.* **72:** 323–327.

Forthofer, R. N., and G. G. Koch. 1973. "An Analysis for Compounded Functions of Categorical Data." *Biometrics* **29:** 143–157.

Forthofer, R. N., and R. G. Lehnen. 1981. *Public Program Analysis, A New Categorical Data Approach.* Belmont, Calif: Lifetime Learning Publications.

Freedman, D., R. Pisani, and R. Purves. 1978. *Statistics.* New York: W. W. Norton.

Freeman, L. C. 1976. "A Further Note on Freeman's Measure of Association." *Psychometrika* **41:** 273–275.

Gans, L. P., and C. A. Robertson. 1981a. "Distributions of Goodman and Kruskal's Gamma and Spearman's Rho in 2 × 2 Tables for Small and Moderate Sample Sizes." *J. Amer. Statist. Assoc.* **76:** 942–946.

Gans, L. P., and C. A. Robertson. 1981b. "The Behavior of Estimated Measures of Association in Small and Moderate Sample Sizes for 2 by 3 Tables." *Commun. Statist.* **A 10:** 1673–1686.

Gart, J. J., and J. R. Zweiful. 1967. "On the Bias of Various Estimators of the Logit and its Variance with Application to Quantal Bioassay." *Biometrika* **54:** 181–187.

Gilbert, G. N. 1981. *Modelling Society.* London: Allen & Unwin.

Gilula, Z. 1982. "A Note on the Analysis of Association in Cross Classifications Having Ordered Categories." *Commun. Statist.* **A 11:** 1233–1240.

Glass, D. V., ed. 1954. *Social Mobility in Britain.* Glencoe, Il: Free Press.

Gokhale, D. V., and S. Kullback. 1978. *The Information in Contingency Tables.* New York: Marcel-Dekker.

Goodman, L. A. 1968. "The Analysis of Cross-Classified Data: Independence, Quasi-Independence, and Interaction in Contingency Tables with or without Missing Cells." *J. Amer. Statist. Assoc.* **63:** 1091–1131.

Goodman, L. A. 1970. "The Multivariate Analysis of Qualitative Data: Interactions among Multiple Classifications" *J. Amer. Statist. Assoc.* **65:** 226–256.

Goodman, L. A. 1971. "The Analysis of Multidimensional Contingency Tables: Stepwise Procedures and Direct Estimation Methods for Building Models for Multiple Classifications." *Technometrics* **13:** 33–61.

Goodman, L. A. 1972. "Some Multiplicative Models for the Analysis of Cross-Classified Data." *Proc. 6th Berkeley Symposium on Mathematical Statistics and Probability.* Vol. **1:** Berkeley: University of California Press, pp. 649–696.

Goodman, L. A. 1978. *Analyzing Qualitative/Categorical Data: Log-Linear Analysis and Latent Structure Analysis.* Cambridge, Mass.: Abt. (Reprints several Goodman articles.)

Goodman, L. A. 1979a. "Simple Models for the Analysis of Association in Cross-Classifications Having Ordered Categories." *J. Amer. Statist. Assoc.* **74:** 537–552.

Goodman, L. A., 1979b. "Multiplicative Models for Square Contingency Tables with Ordered Categories." *Biometrika* **66:** 413–418.

Goodman, L. A. 1981a. "Association Models and Canonical Correlation in the Analysis of Cross-Classifications Having Ordered Categories." *J. Amer. Statist. Assoc.* **76:** 320–334.

Goodman, L. A. 1981b. "Association Models and the Bivariate Normal Distribution in the Analysis of Cross-Classifications Having Ordered Categories." *Biometrika* **68:** 347–355.

Goodman, L. A. 1981c. "Three Elementary Views of Loglinear Models for the Analysis of Cross-Classifications Having Ordered Categories." *Sociological Methodology*. San Francisco: Jossey-Bass, pp. 193–239.

Goodman, L. A. 1983. "The Analysis of Dependence in Cross-Classifications Having Ordered Categories, Using Log-Linear Models for Frequencies and Log-Linear Models for Odds." *Biometrics* **39:** 149–160.

Goodman, L. A., and W. H. Kruskal. 1954. "Measures of Association for Cross Classifications." *J. Amer. Statist. Assoc.* **49:** 732–764.

Goodman, L. A., and W. H. Kruskal. 1959. "Measures of Association for Cross Classifications II: Further Discussion and References." *J. Amer. Statist. Assoc.* **54:** 123–163.

Goodman, L. A., and W. H. Kruskal. 1963. "Measures of Association for Cross Classifications III: Approximate Sampling Theory." *J. Amer. Statist. Assoc.* **58:** 310–364.

Goodman, L. A., and W. H. Kruskal. 1972. "Measures of Association for Cross Classifications IV: Simplification of Asymptotic Variances." *J. Amer. Statist. Assoc.* **67:** 415–421.

Goodman, L. A., and W. H. Kruskal. 1979. *Measures of Association for Cross Classifications*. New York: Springer-Verlag. (Contains four articles just listed.)

Grizzle, J. E., C. F. Starmer, and G. G. Koch. 1969. "Analysis of Categorical Data by Linear Models." *Biometrics* **25:** 489–504.

Gross, S. T. 1981. "On Asymptotic Power and Efficiency of Tests of Independence in Contingency Tables with Ordered Classifications." *J. Amer. Statist. Assoc.* **76:** 935–941.

Grove, D. M. 1980. "A Test of Independence against a Class of Ordered Alternatives in a $2 \times c$ Contingency Table." *J. Amer. Statist. Assoc.* **75:** 454–459.

Gurland, J., I. Lee, and P. A. Dahm. 1960. "Polychotomous Quantal Response in Biological Assay." *Biometrics* **16:** 382–398.

Haberman, S. J. 1973. "The Analysis of Residuals in Cross-Classified Tables." *Biometrics* **29:** 205–220.

Haberman, S. J. 1974a. *The Analysis of Frequency Data*. Chicago: University of Chicago Press.

Haberman, S. J. 1974b. "Loglinear Models for Frequency Tables with Ordered Classifications." *Biometrics* **30:** 589–600.

Haberman, S. J. 1978. *Analysis of Qualitative Data. Vol. 1: Introductory Topics*. New York: Academic.

Haberman, S. J. 1979. *Analysis of Qualitative Data. Vol. 2: New Developments*. New York: Academic.

Haberman, S. J. 1981. "Tests for Independence in Two-Way Contingency Tables Based on Canonical Correlation and on Linear-by-Linear Interaction." *Ann. Statist.* **9:** 1178–1186.

Hastie, T. J., and J. M. Juritz. 1981. "The Stability of Ordinal Measures of Association in Contingency Tables." Unpublished manuscript.

Hawkes, R. K. 1971. "The Multivariate Analysis of Ordinal Measures." *Amer. J. Sociol.* **76:** 908–926.

Hawkes, R. K. 1976. "The Effects of Grouping on Measures of Ordinal Association." *Sociological Methodology*. San Francisco: Jossey-Bass, pp. 176–194.

Hedlund, R. D. 1978. "Cross-Over Voting in a 1976 Open Presidential Primary." *Public Opinion Quart.* **41:** 498–514.

Hildebrand, D. K., J. D. Laing, and H. Rosenthal. 1977a. *Prediction Analysis of Cross Classifications*. New York: Wiley.

Hildebrand, D. K., J. D. Laing, and H. Rosenthal. 1977b. *Analysis of Ordinal Data*. Sage

University Paper Series on Quantitative Applications in the Social Sciences. Series no. 07-008. Beverly Hills and London: Sage Publications.

Holmes, M. C., and R. E. O. Williams, 1954. "The Distribution of Carriers of *Streptococcus pyogenes* among 2413 Healthy Children." *J. Hyg. Camb.* **52:** 165–179.

Imrey, P. B., G. G. Koch, and M. E. Stokes. 1982. "Categorical Data Analysis: Some Reflections on the Log Linear Model and Logistic Regression. Part II: Data Analysis." *Intern. Statist. Rev.* **50:** 35–63.

Jørgensen, B. 1983. "Maximum Likelihood Estimation and Large-Sample Inference for Generalized Linear and Nonlinear Regression Models." *Biometrika* **70:** 19–28.

Kendall, M. G. 1938. "A New Measure of Rank Correlation." *Biometrika* **30:** 81–93.

Kendall, M. G. 1945. "The Treatment of Ties in Rank Problems." *Biometrika* **33:** 239–251.

Kendall, M. G. 1970. *Rank Correlation Methods.* 4th ed. London: Griffin.

Kendall, M. G., and A. Stuart. 1973. *The Advanced Theory of Statistics.* Vol. 2. 2d ed. New York: Hafner.

Kim, J. 1971. "Predictive Measures of Ordinal Association." *Amer. J. Sociol.* **76:** 891–907.

Kim, J. 1975. "Multivariate Analysis of Ordinal Variables." *Amer. J. Sociol.* **81:** 261–298.

Kim, J. 1978. "Multivariate Analysis of Ordinal Variables Revisited." *Amer. J. Sociol.* **84:** 448–456.

Klotz, J. 1980. "A Modified Cochran-Friedman Test with Missing Observations and Ordered Categorical Data." *Biometrics* **36:** 665–670.

Klotz, J., and J. Teng. 1977. "One-Way Layout for Counts and the Exact Enumeration of the Kruskal-Wallis H Distribution with Ties." *J. Amer. Statist. Assoc.* **72:** 165–169.

Knoke, D., and P. J. Burke. 1980. *Log-Linear Models.* Sage University Paper Series on Quantitative Applications in the Social Sciences. Series no. 07-020. Beverly Hills and London: Sage publications.

Koch, G. G., J. L. Freeman, and R. G. Lehnen. 1976. "A General Methodology for the Analysis of Ranked Policy Preference Data." *Intern. Statist. Rev.* **44:** 1–28.

Koch, G. G., D. B. Gillings, and M. E. Stokes. 1980. "Biostatistical Implications of Design, Sampling, and Measurement to Health Science Data Analysis." *Ann. Rev. Public Health* **1:** 163–225.

Koch, G. G., I. A. Amara, G. W. Davis, and D. B. Gillings. 1981. "A Review of Some Statistical Methods for Covariance Analysis." *Biometrics* **38:** 563–595.

Koehler, K., and K. Larntz. 1980. "An Empirical Investigation of Goodness-of-Fit Statistics for Sparse Multinomials." *J. Amer. Statist. Assoc.* **75:** 336–344.

Kruskal, W. H. 1958. "Ordinal Measures of Association." *J. Amer. Statist. Assoc.* **53:** 814–861.

Ku, H. H., and S. Kullback. 1974. "Loglinear Models in Contingency Table Analysis." *Amer. Statist.* **28:** 115–122.

Labovitz, S. 1970. "The Assignment of Numbers to Rank Order Categories." *Amer. Sociol. Rev.* **35:** 515–524.

Laird, N. 1979. "A Note on Classifying Ordinal-Scale Data." *Sociological Methodology.* San Francisco: Jossey-Bass, pp. 303–310.

Lancaster, H. O., and M. A. Hamdan. 1964. "Estimation of the Correlation Coefficient in Contingency Tables with Possibly Nonmetrical Characters." *Psychometrika* **29:** 383–391.

Landis, J. R., W. M. Stanish, J. L. Freeman, and G. G. Koch. 1976. "A Computer Program for the Generalized Chi-Square Analysis of Categorical Data Using Least Squares." *Comput. Programs Biomed.* **6:** 196–231.

Landis, J. R., M. M. Cooper, T. Kennedy, and G. G. Koch. 1978. "A Computer Program for Testing Average Partial Association in Three-Way Contingency Tables (PARCAT)." *Comput. Programs Biomed.* **9:** 223–246.

Landis, J. R., E. R. Heyman, and G. G. Koch. 1978. "Average Partial Association in Three-Way Contingency Tables: A Review and Discussion of Alternative Tests." *Intern. Statist. Rev.* **46:** 237–254.

Larntz, K. 1978. "Small-Sample Comparison of Exact Levels for Chi-Squared Goodness-of-Fit Statistics." *J. Amer. Statist. Assoc.* **73:** 253–263.

Lehmann, E. L. 1966. "Some Concepts of Dependence." *Ann. Math. Statist.* **37:** 1137–1153.

Lehmann, E. L. 1975. *Nonparametrics: Statistical Methods Based on Ranks.* San Francisco: Holden-Day.

Leik, R. K. 1976. "Monotone Regression Analysis for Ordinal Variables." *Sociological Methodology.* San Francisco: Jossey-Bass, pp. 250–270.

Levin, B., and P. E. Shrout. 1981. "On Extending Bock's Model of Logistic Regression in the Analysis of Categorical Data." *Commun. Statist.* **A 10:** 125–147.

Lindgren, B. W. 1976. *Statistical Theory.* 3d ed. New York: Macmillan.

Madsen, M. 1976. "Statistical Analysis of Multiple Contingency Tables: Two Examples." *Scand. J. Statist.* **3:** 97–106.

Mantel, N. 1963. "Chi-Squared Tests with One Degree of Freedom; Extensions of the Mantel-Haenszel Procedure." *J. Amer. Statist. Assoc.* **58:** 690–700.

Mantel, N., and W. Haenszel. 1959. "Statistical Aspects of the Analysis of Data from Retrospective Studies of Disease." *J. Natl. Cancer Inst.* **22:** 719–748.

Mayer, L. S. 1971. "A Note on Treating Ordinal Data as Interval Data." *Amer. Sociol. Rev.* **36:** 449–453.

Mayer, L. S., and J. A. Robinson. 1978. "Measures of Association for Multiple Regression Models with Ordinal Predictor Variables." *Sociological Methodology.* San Francisco: Jossey-Bass, pp. 141–163.

McCullagh, P. 1977. "A Logistic Model for Paired Comparisons with Ordered Categorical Data." *Biometrika* **64:** 449–453.

McCullagh, P. 1978. "A Class of Parametric Models for the Analysis of Square Contingency Tables with Ordered Categories." *Biometrika* **65:** 413–418.

McCullagh, P. 1979. "The Use of the Logistic Function in the Analysis of Ordinal Data." *Proc. Internat. Statist. Inst.*, Manila.

McCullagh, P. 1980. "Regression Models for Ordinal Data" (with discussion). *J. Roy. Statist. Soc.* **B 42:** 109–142.

McCullagh, P. 1982. "Some Applications of Quasisymmetry." *Biometrika* **69:** 303–308.

McCullagh, P., and J. Nelder. 1983. *Generalized Linear Models.* London: Chapman and Hall.

McKelvey, R. D., and W. Zavoina. 1975. "A Statistical Model for the Analysis of Ordinal Level Dependent Variables." *J. Mathematical Sociol.* **4:** 103–120.

Mohberg, N. R., M. Ghosh, and J. E. Grizzle. 1978. "Linear Models Analysis of Small Samples of Categorized Ordinal Response Data." *J. Statist. Comp. Simul.* **7:** 237–252.

Mosteller, F. 1968. "Association and Estimation in Contingency Tables." *J. Amer. Statist. Assoc.* **63:** 1–28.

Nelder, J. A., and R. W. M. Wedderburn. 1972. "Generalized Linear Models." *J. Roy. Statist. Soc.* **A 135:** 370–384.

Neter, J., and W. Wasserman. 1974. *Applied Linear Statistical Models.* Homewood, Il: Irwin.

Nie, N. H., C. H. Hull, J. G. Jenkins, K. Steinbrenner, and D. H. Bent. 1975. *Statistical Package for the Social Sciences.* New York: McGraw-Hill.

O'Brien, R. M. 1980. "The Use of Pearson's *r* with Ordinal Data." *Amer. Sociol. Rev.* **45:** 851–857.

Palmgren, J. 1981. "The Fisher Information Matrix for Log Linear Models Arguing Conditionally on Observed Explanatory Variables." *Biometrika* **68:** 563–566.

Patefield, W. M. 1982. "Exact Tests for Trends in Ordered Contingency Tables." *J. Roy. Statist. Soc.* **C 31:** 32–43.

Pearson, E. S., and H. O. Hartley. 1972. *Biometrika Tables for Statisticians, Vol II.* Cambridge: Cambridge University Press.

Plackett, R. L. 1965. "A Class of Bivariate Distributions." *J. Amer. Statist. Assoc.* **60:** 516–522.

Plackett, R. L. 1981. *The Analysis of Categorical Data.* 2d ed. London: Griffin.

Plackett, R. L., and S. R. Paul. 1978. "Dirichlet Models for Square Contingency Tables." *Commun. Statist.* **A 7:** 939–952.

Ploch, D. R. 1974. "Ordinal Measures of Association and the General Linear Model." Chapter 12 in *Measurement in the Social Sciences.* Ed. by H. M. Blalock. Chicago: Aldine.

Pratt, J. W. 1981. "Concavity of the Log Likelihood." *J. Amer. Statist. Assoc.* **76:** 103–106.

Prentice, R. L., and L. A. Gloeckler. 1978. "Regression Analysis of Grouped Survival Data with Application to Breast Cancer Data." *Biometrics* **34:** 57–67.

Proctor, C. H. 1973. "Relative Efficiencies of Some Measures of Association for Ordered Two-Way Contingency Tables under Varying Intervalness of Measurement Error." *Proc. Soc. Statist. Sec., ASA*, pp. 372–379.

Quade, D. 1974. "Nonparametric Partial Correlation." Chapter 13 in *Measurement in the Social Sciences.* Ed. by H. M. Blalock. Chicago: Aldine.

Radelet, M. 1981. "Racial Characteristics and the Imposition of the Death Penalty." *Amer. Sociol. Rev.* **46:** 918–927.

Radelet, M., and G. L. Pierce. 1983. "Race and Prosecutorial Discretion in Homicide Cases." Paper presented at annual meeting of Amer. Sociol. Assoc., Detroit.

Rao, C. R. 1973. *Linear Statistical Inference and Its Applications.* 2d ed. New York: Wiley.

Ray, A. A. 1982. *SAS User's Guide: Statistics*, Cary, N.C.: SAS Institute, Inc.

Reynolds, H. T. 1974. "Ordinal Partial Association and Causal Inferences." Chapter 14 in *Measurement in the Social Sciences.* Ed. by H. M. Blalock. Chicago: Aldine.

Reynolds, H. T. (1977). *The Analysis of Cross-Classifications.* New York: Free Press.

Robertson, T., and F. T. Wright. 1981. "Likelihood Ratio Tests for and against a Stochastic Ordering Between Multinomial Populations." *Ann. Statist.* **9:** 1248–1257.

Rosenthal, I. 1966. "Distribution of the Sample Version of the Measure of Association, Gamma." *J. Amer. Statist. Assoc.* **61:** 440–453.

Schollenberger, J., A. Agresti, and D. Wackerly. 1979. "Measuring Association and Modelling Relationships Between Interval and Ordinal Variables." *Proc. Soc. Statist. Sec., ASA*, pp. 624–626.

Semenya, K., and G. G. Koch, 1979. "Linear Models Analysis for Rank Functions of Ordinal Categorical Data." *Proc. Statist. Comp. Sec., ASA*, pp. 271–276.

Semenya, K., and G. G. Koch. 1980. "Compound Function and Linear Model Methods for the Multivariate Analysis of Ordinal Categorical Data." Mimeo Series No. 1323. Chapel Hill: University of North Carolina Institute of Statistics.

Semenya, K., G. G. Koch, M. E. Stokes, and R. N. Forthofer. 1983. "Linear Models Methods for Some Rank Function Analyses of Ordinal Categorical Data." *Commun. Statist.* **A 12:** 1277–1298.

Simon, G. A. 1974. "Alternative Analyses for the Singly-Ordered Contingency Table." *J. Amer. Statist. Assoc.* **69:** 971–976.

Simon, G. A. 1978. "Efficacies of Measures of Association for Ordinal Contingency Tables." *J. Amer. Statist. Assoc.* **73:** 545–551.

Simpson, E. H. 1951. "The Interpretation of Interaction in Contingency Tables." *J. Roy. Statist. Soc.* **B 13:** 238–241.

Smith, K. W. 1976. "Marginal Standardization and Table Shrinking: Aids in the Traditional Analysis of Contingency Tables." *Social Forces* **54:** 669–693.

Snell, E. J. 1964. "A Scaling Procedure for Ordered Categorical Data." *Biometrics* **20:** 592–607.

Somers, R. H. 1959. "The Rank Analogue of Product Moment Partial Correlation and Regression, with Application to Manifold, Ordered Contingency Tables." *Biometrika* **46:** 241–246.

Somers, R. H. 1962. "A New Asymmetric Measure of Association for Ordinal Variables." *Amer. Sociol. Rev.* **27:** 799–811.

Somers, R. H. 1968. "An Approach to the Multivariate Analysis of Ordinal Data." *Amer. Sociol. Rev.* **33:** 971–977.

Somers, R. H. 1974. "Analysis of Partial Rank Correlation Measures Based on the Product-Moment Model: Part One." *Social Forces* **53:** 229–246.

Srole, L., T. S. Langner, S. T. Michael, P. Kirkpatrick, M. K. Opler, and T. A. C. Rennie. 1978. *Mental Health in the Metropolis: The Midtown Manhattan Study.* Rev. ed. New York: NYU Press.

Stuart, A. 1953. "The Estimation and Comparison of Strengths of Association in Contingency Tables." *Biometrika* **40:** 105–110.

Stuart, A. 1955. "A Test for Homogeneity of the Marginal Distributions in a Two-Way Classification." *Biometrika* **42:** 412–416.

Stuart, A. 1963. "Calculation of Spearman's Rho for Ordered Two-Way Classifications." *Amer. Statist.* **17:** 23–24.

Theil, H. 1970. "On the Estimation of Relationships Involving Qualitative Variables." *Amer. J. Sociol.* **76:** 103–154.

Thompson, W. A. 1977. "On the Treatment of Grouped Observations in Life Studies." *Biometrics* **33:** 463–470.

Upton, G. J. G. 1978. *The Analysis of Cross-Tabulated Data.* New York: Wiley.

Vigderhous, G. 1979. "Equivalence between Ordinal Measures of Association and Tests of Significant Differences between Samples." *Quality and Quantity* **13:** 187–201.

Wahrendorf, J. 1980. "Inference in Contingency Tables with Ordered Categories Using Plackett's Coefficient of Association for Bivariate Distributions." *Biometrika* **67:** 15–21.

Walker, S. H., and D. B. Duncan. 1967. "Estimation of the Probability of an Event as a Function of Several Independent Variables." *Biometrika* **54:** 167–179.

Whittaker, J., and M. Aitkin. 1978. "A Flexible Strategy for Fitting Complex Log-Linear Models." *Biometrics* **34:** 487–495.

Whittemore, A. S. 1978. "Collapsibility of Multidimensional Contingency Tables." *J. Roy. Statist. Soc.* **B 40:** 328–340.

Williams, E. J. 1952. "Use of Scores for the Analysis of Association in Contingency Tables." *Biometrika* **39**: 274–289.

Williams, O. D., and J. E. Grizzle. 1972. "Analysis of Contingency Tables Having Ordered Response Categories." *J. Amer. Statist. Assoc.* **67**: 55–63.

Wilson, T. P. 1969. "A Proportional Reduction in Error Interpretation for Kendall's Tau-*b*." *Social Forces* **47**: 340–342.

Yates, F. 1934. "Contingency Tables Involving Small Numbers and the χ^2 Test." *J. Roy. Statist. Soc.*, supp. 1, 217–235.

Yates, F. 1948. "The Analysis of Contingency Tables with Groupings Based on Quantitative Characters." *Biometrika* **35**: 176–181.

Yule, G. U. 1903. "Notes on the Theory of Association of Attributes in Statistics." *Biometrika* **2**: 121–134.

Yule, G. U. 1912. "On the Methods of Measuring Association between Two Attributes." *J. Roy. Statist. Soc.* **75**: 579–642.

Index of Examples

Index of Selected Notation

Note: page number indicates first appearance in text.

Author Index

Subject Index

Applied Probability and Statistics (Continued)

DRAPER and SMITH • Applied Regression Analysis, *Second Edition*
DUNN • Basic Statistics: A Primer for the Biomedical Sciences, *Second Edition*
DUNN and CLARK • Applied Statistics: Analysis of Variance and Regression
ELANDT-JOHNSON and JOHNSON • Survival Models and Data Analysis
FLEISS • Statistical Methods for Rates and Proportions, *Second Edition*
FOX • Linear Statistical Models and Related Methods
FRANKEN, KÖNIG, ARNDT, and SCHMIDT • Queues and Point Processes
GALAMBOS • The Asymptotic Theory of Extreme Order Statistics
GIBBONS, OLKIN, and SOBEL • Selecting and Ordering Populations: A New Statistical Methodology
GNANADESIKAN • Methods for Statistical Data Analysis of Multivariate Observations
GOLDBERGER • Econometric Theory
GOLDSTEIN and DILLON • Discrete Discriminant Analysis
GREENBERG and WEBSTER • Advanced Econometrics: A Bridge to the Literature
GROSS and CLARK • Survival Distributions: Reliability Applications in the Biomedical Sciences
GROSS and HARRIS • Fundamentals of Queueing Theory
GUPTA and PANCHAPAKESAN • Multiple Decision Procedures: Theory and Methodology of Selecting and Ranking Populations
GUTTMAN, WILKS, and HUNTER • Introductory Engineering Statistics, *Third Edition*
HAHN and SHAPIRO • Statistical Models in Engineering
HALD • Statistical Tables and Formulas
HALD • Statistical Theory with Engineering Applications
HAND • Discrimination and Classification
HILDEBRAND, LAING, and ROSENTHAL • Prediction Analysis of Cross Classifications
HOAGLIN, MOSTELLER, and TUKEY • Understanding Robust and Exploratory Data Analysis
HOEL • Elementary Statistics, *Fourth Edition*
HOEL and JESSEN • Basic Statistics for Business and Economics, *Third Edition*
HOGG and KLUGMAN • Loss Distributions
HOLLANDER and WOLFE • Nonparametric Statistical Methods
IMAN and CONOVER • Modern Business Statistics
JAGERS • Branching Processes with Biological Applications
JESSEN • Statistical Survey Techniques
JOHNSON and KOTZ • Distributions in Statistics
Discrete Distributions
Continuous Univariate Distributions—1
Continuous Univariate Distributions—2
Continuous Multivariate Distributions
JOHNSON and KOTZ • Urn Models and Their Application: An Approach to Modern Discrete Probability Theory
JOHNSON and LEONE • Statistics and Experimental Design in Engineering and the Physical Sciences, Volumes I and II, *Second Edition*
JUDGE, HILL, GRIFFITHS, LÜTKEPOHL and LEE • Introduction to the Theory and Practice of Econometrics
JUDGE, GRIFFITHS, HILL and LEE • The Theory and Practice of Econometrics
KALBFLEISCH and PRENTICE • The Statistical Analysis of Failure Time Data